KU-340-921

OCCUPATIONAL HYGIENE

Occupational Hygiene

An Introductory Guide

ALAN L. JONES, DAVID M.W. HUTCHESON AND SARAH M. DYMOTT

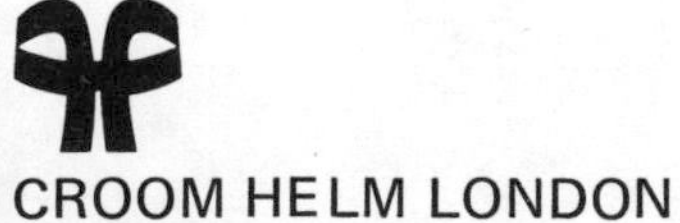

CROOM HELM LONDON

Croom Helm Ltd, 2-10 St John's Road, London SW11

British Library Cataloguing in Publication Data

Jones, Alan L.
Occupational hygiene.
1. Industrial hygiene
I. Title II. Hutcheson, David M.W.
III. Dymott, Sarah M.
613.6'2024613 R967

ISBN 0-7099-1404-0

Printed and bound in Great Britain by
Biddles Ltd, Guildford and King's Lynn

CONTENTS

ACKNOWLEDGEMENTS

The authors would like to thank the many colleagues in the occupational health field who gave valuable, constructive comments on the manuscript of this book. The vast majority of these comments have been incorporated in the text and have served to increase not only the breadth of the subjects but, hopefully, also the depth of the reader's understanding.

We would also like to express our thanks to Julia for her constant patience during the seemingly interminable manuscript preparations.

PREFACE

The late and much missed Dr Suzette Gauvain preached and practised the team approach to occupational health (OH). Her conviction was that OH could not work effectively if the people involved did not work as a team. As a result, the last years of her life were spent in attempting to pull together the doctors, hygienists and OH nurses so that each could understand the other and could contribute effectively – even outside their own sphere of learning.

This book is, in essence, one of the fruits of Dr Gauvain's movement. The truth of the effectiveness of the team approach is, in fact, soon understood, but that is the way of all inventions – it becomes obvious once someone else has already said it. Let us not forget the pioneers however, since ultimately we follow the paths laid down by them – even if we believe our contribution is the major one.

'Contribution' is, in fact, the key word in defining the team approach – no matter what the objective of the team. This is because a group of people, working with a joint objective, must involve the opinions and expertise of each member of the team, if the objective is to be reached with the best possible solution.

One of the reasons that OH teams have failed in the past is the lack of understanding that each member has of the others' area of expertise. Attempts have been made to increase the medical knowledge of hygienists and the hygiene knowledge of doctors. Indeed, several courses are run whereby doctors and hygienists are taught in the same group, e.g. the London School of Hygiene and Tropical Medicine, Master of Science Degrees in Occupational Hygiene and Occupational Medicine, which attempt to bring the doctors and hygienists together where at all possible. Training of OH nurses, however, has presented problems – particularly in the hygiene field – because of the lack of texts aimed specifically at covering the topic for this audience.

The purpose of this book, therefore, is to provide the necessary general information that will relate to the sort of training OH nurses receive. It will also provide safety representatives with sufficient knowledge to enable more involvement in this field. Since 1978, the number of safety representatives and their involvement in OH has increased enormously. Training is via TUC courses or in-house company schemes. We believe that this book will provide these people with a reference text that will enable the safety representatives to contribute more in the hygiene field than has proven possible to date.

Each chapter has been written in a manner which will allow the reader to gain the fundamental concepts in each of the fields. The book also aims to provide sufficient guidance and references so that where a specific need arises, the reader can obtain direction to a specialist text for further information. This is all the more important since large numbers of OH nurses in particular work completely alone in industry.

Occupational hygiene is a relatively young field. Indeed, it was only in 1700 that Bernardino Ramazzini – the father of occupational medicine – published his observations of occupational diseases in *De Morbis Artificum Diatriba* (Diseases of Tradesmen). There is undoubtedly a great deal yet to be achieved. Industrial diseases are still as common as they are unacceptable. The solution to this situation is not simply enforcement of legislation (currently a physically impossible task) but the production of occupational health teams that can provide management with acceptable changes that enable the job to be done as efficiently but with minimum hazard.

It is hoped that the reader will enjoy this text, for occupational hygiene is a developing science and as such should be both challenging and enjoyable.

OCCUPATIONAL HYGIENE

INTRODUCTION

The ideal occupational health team is made up of occupational health doctors, nurses, hygienists, safety officers and now, no less important, the new group to join the team, the safety representatives. The last are much welcomed.

It is now understood that to function effectively the team approach is vital in a comprehensive occupational health programme. All the particular specialities in the occupational health team overlap their roles and knowledge of each member's work is absolutely essential. This cannot be better emphasised than by examining the agreed objectives of an occupational health programme.

Objectives of an Occupational Health Programme

The joint ILO/WHO Committee on Occupational Health 1950, the ILO's recommendation No. 112 (1959), which was endorsed in a Recommendation of the EEC Commission in 1962 and in a Resolution of the Committee of Ministers of the Council of Europe in 1972, defined the purposes of an occupational health service as:

> The promotion and maintenance of the highest possible level of health among the gainfully employed upon whom the economic welfare of the community depends, is the main objective of an Occupational Health programme.

To meet these objectives it is necessary:

(1) to identify and bring under control all the *chemical, physical, mechanical, biological* and *psychosocial* agents that are known or thought to be hazardous;
(2) to ensure that the physical and mental demands on those at work and the jobs they do are matched with their *individual anatomical, physiological* and *psychological* capabilities, needs and limitations;
(3) to provide effective measures to protect those who are vulnerable and to raise their level of resistance;

(4) to discover and improve work situations that may contribute to the overall ill-health of workers;
(5) to educate management and the workforce to fulfil their responsibilities relevant to health protection and promotion;
(6) to carry out comprehensive in-plant health programmes dealing with men's total health, which will assist public health authorities to raise the level of community health.

The occupational health service in complying with these objectives can so identify and solve problems of health related to work.

The function of the industrial hygienist in this unique field includes:

(1) examination of the industrial environment for potential health hazards;
(2) interpretation of the data from studies in the industrial environment;
(3) preparation of control measures and their assessment;
(4) assisting in the setting of standards for work conditions;
(5) presentation of competent facts;
(6) preparation of adequate warnings and precautions where danger exists;
(7) education of the workforce in occupational hygiene;
(8) contribution to studies of health effects, e.g. epidemiological studies.

It is neither possible nor desirable for each team member to be an 'expert' in all aspects of the team's work. However, the guides to hygiene set out in this book should provide sufficient information to allow the other team members to contribute significantly to this field.

1 BASIC PHYSIOLOGY

Introduction

Occupational hygiene is the study of the evaluation and control of workplace environments potentially detrimental to the health and well-being of the worker. Before the reader can gain insight into hygiene activities however, he or she must have a certain knowledge of the 'target' organs of the body that can be attacked by materials present in occupational environments. The purpose of this introductory chapter therefore, is to provide a very basic understanding of the key 'target' organs important from an industrial hygiene viewpoint. Each target organ is discussed also in the context of what chemicals or physical agents are most likely to produce a change in the function of the organ. In the case of the section on the lung, a brief description is also given of the function tests that can be carried out in order to monitor for signs of impairment – since this is one of the key problem areas in occupational health.

Obviously this chapter is meant mainly for readers other than nurses since the majority of the information will already have been covered in greater depth in the nurse's standard training.

Target Organ 1: The Lungs

The lungs are the main organs of the respiratory system which can be considered to be divided into an *upper* respiratory system (nose, pharynx and neck) and a *lower* respiratory system (trachea, tracheobronchial system and lungs).

The function of this system is to provide the means for air to come into intimate contact with the blood so that oxygen can be transferred from the air into the blood and 'waste' carbon dioxide can be transferred into the air and out of the body. The lungs should therefore be regarded as *external organs* in that they are *directly* in contact with the outside environment. A simplified drawing (Figure 1.1) provides some guide as to various parts of the airways in the respiratory system.

The body requires about 250 cm^3 of oxygen every minute for basic metabolism (converting food into heat and energy) when the body is at rest. When very severe work is being carried out this 'need' is increased, sometimes to as high as 5000 cm^3 of oxygen per minute. Carbon dioxide is formed when the food reacts with the oxygen to give heat and energy, and this must be removed via the lungs since it is harmful. The air in the alveoli therefore contains a larger amount of carbon dioxide than in normal room air and a lower amount of oxygen, i.e. alveoli air contents is 14 per cent oxygen and 5.5-6 per cent carbon dioxide, whereas normal room air contains 21 per cent oxygen and 0.04 per cent carbon dioxide.

Figure 1.1: Respiratory Airways

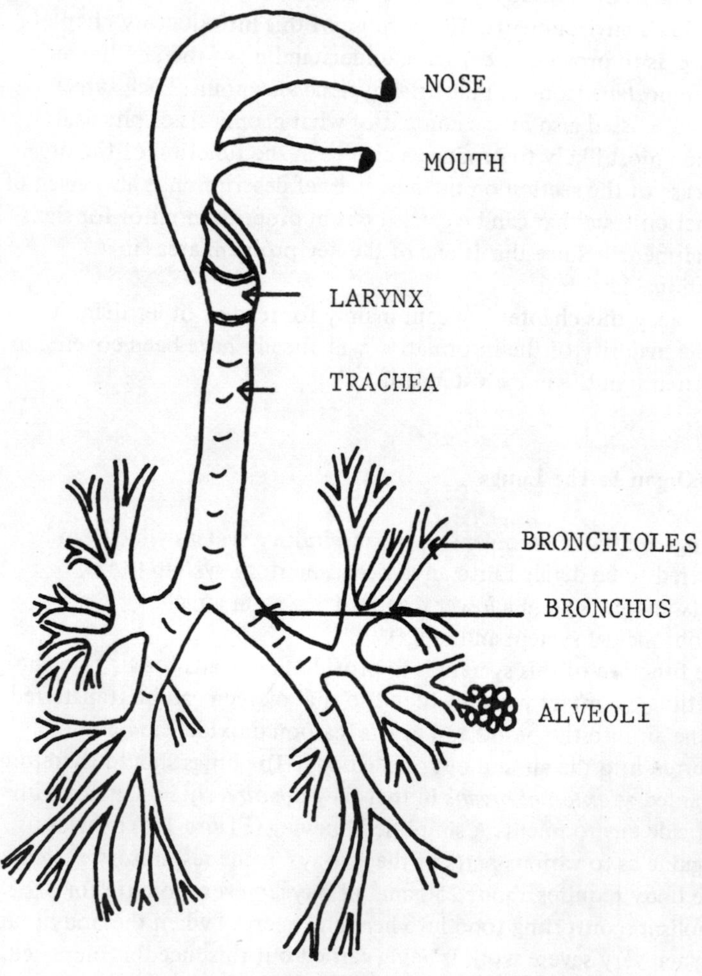

At the end of a normal breath expiration (after breathing out) the lungs still contain about three litres of air and this is just about enough to keep a person supplied with oxygen for about two minutes before more oxygen is needed.

The amount of air breathed in is obviously important in controlling how much oxygen reaches the alveoli. However, just as important is how many alveoli are available and working properly so that the oxygen can be fully utilised. These two factors are the major 'functions' of the respiratory system that can be impaired by certain disorders. For example, if a worker undergoes lung function tests which show that when he or she forces the air out of the lungs as quickly as possible and the time taken to do this is much longer than normal, then an *obstructive* defect can be diagnosed – i.e. the tubes taking the air to the alveoli are partially blocked. This can be due to an asthmatic condition (the tubes constrict) which can be caused by an allergy to a workplace (or home) contaminant. Bronchitis has the same effect, but in this case the obstruction is due to 'mucous' blocking the tubes.

If it is found that a worker's lung function tests indicate a much smaller lung volume when forcibly exhaling, then a *restrictive* defect can be diagnosed. This can be caused by the alveoli losing the 'elasticity' necessary to allow air in and push it out. Causes of this can be asbestosis (caused by inhaling asbestos fibres) or silicosis (caused by inhaling crystalline silica). These are typified by the fact that they cause the alveoli walls to become fibrotic and 'harden' so that they lack the flexibility to expand and contract whilst air is breathed in and out respectively.

The lung function tests available are carried out by the OH nurse – therefore no details are given here, except to list the usual ones.

(1) *Spirometer.* This measures the lung volumes whilst the subject is breathing normally and these can be traced on paper (by means of a pen recorder) and examined later.

(2) *Vitallograph.* This measures the forced lung volumes, i.e. the subject is asked to take a deep breath and told to exhale through the vitallograph as quickly and as hard as possible. The exhaled breath is measured on a time scale so that the rate at which the air is exhaled is measured and a standard based on the amount of air forced out during the first second (Forced Expiratory Volume in one second – FEV_1) compared to the *total* amount of air forced out (Forced Vital Capacity – FVC).

(3) *Pneumotachograph.* This is used to measure the condition of the *small* airways nearest to the alveoli.

(4) *Others.* The transfer of oxygen across the alveoli into the blood can be measured using various tests which require skilled operators and are not usually measured as a *routine* health check.

Target Organ 2: The Liver

The liver is the 'chemical plant' of the body and is vital for a number of reasons:

(1) It is the major detoxifying organ in the body, e.g. protein break-down products are converted into urea which is excreted via the urine. Foreign chemical substances are also attacked in the liver, with the object of making them more soluble in water so that they can be excreted in the urine.
(2) It produces plasma proteins for the blood.
(3) It produces heparin which is the material which causes blood to clot.
(4) It regulates the metabolism/intake of glucose sugar to maintain a fixed level in the body.
(5) It produces bile which is necessary to absorb fats from the diet.

Because of its function, the liver has a large supply of blood both from the lungs (hepatic artery carrying oxygenated blood) and from the guts (portal vein carrying blood used to absorb 'material' from the guts). For this reason most chemicals alien to the body will at some time end up in the liver for conversion into a form that can be removed from the body. Knowledge of what the action of the liver is on each chemical can therefore help to gauge the absorption of that chemical by measuring the level of the transformed chemical (metabolite) in the urine (or the faeces and blood).

Examples of these are:

(1) *Ethyl alcohol* is converted in the liver to acetaldehyde which in turn is converted to a chemical used in one of the major heat/energy producing reactions. Because of this, unused fat (normally used for heat/energy) builds up around the liver and a 'fatty' liver is an early warning sign of liver damage.

(2) *Cyanide* (highly toxic) compounds are converted by the liver into thiocyanates which can be removed in the urine.
(3) *Benzene* is converted to phenol and then to more soluble materials.

Excessive exposure to toxic chemicals will eventually result in the inability of the liver to handle its numerous tasks efficiently and symptoms will occur depending on which part of the system breaks down. Complete failure of the liver will result in death.

Target Organ 3: The Kidneys

The body has two kidneys. Their prime function is to produce urine and in so doing they:

(1) maintain a water balance;
(2) maintain an electrolyte balance (salt);
(3) maintain the blood pH (acidity/alkalinity).

The working unit in the kidney is called the *nephron* and there are about one million nephrons per kidney. Blood capillaries are linked in a bundle to each nephron at the end of a thin tube called a *tubule.* This 'bundle-end' of the tubule is called the *glomerulus.* The water, electrolytes, waste chemicals, etc., pass through from the blood via the glomerulus and go through the tubule to a main collecting-duct which services several tubules. Each main collecting-duct then leads to a larger duct called the *ureter* which goes direct to the *bladder.* The bladder is periodically voided via the *urethra.*

The glomerulus of course is the 'filter' unit, but if too much water, salt, etc. is allowed to go through, it can be reabsorbed back into the bloodstream via the walls of the tubule. This is therefore one of the body's control mechanisms.

All of the nephrons working together can 'filter' about 120 cm^3 of water per minute, but the tubules will reabsorb about 119 cm^3 per minute, so that only 1 cm^3 per minute will go to the bladder. This is equivalent to about 1500 cm^3 per day of urine (normal).

The effect of 'alien' chemicals can be either to cause blocking of the glomerulus filter or to change the characteristics of the walls of the tubules so that the reabsorption stage is either too high or too low. No symptoms will occur until the glomerulus filtration rate falls from

120 cm^3 down to 30 cm^3 per minute, i.e. down to half a functional kidney, but below that special non-protein diets are required. Examination of the urine in patients suffering from kidney defects will usually help in specifying the cause. Normal urine will contain no glucose, protein or blood; their presence in the urine is therefore an indication of damage. The quantity of urine passed per day is also an important guideline.

Excessive exposure to inorganic mercury compounds will cause the death of the tubule wall cells and also damage the glomerulus, resulting in the 'filter' allowing through larger molecules than normal and the urine will be found to contain protein.

Chronic absorption of cadmium can also cause a nephrotoxic action (toxic effect on the kidneys). When the cadmium concentration is greater than about 200 mg/kg weight, a toxic effect is directed towards the tubules and this can be seen by the fact that protein is again excreted in the urine.

Target Organ 4: The Blood

An adult has about 8-9 pints (5 litres) of blood. Made in some of the *bone marrow* (called red bone marrow) blood consists of about 45 per cent by volume of cells (white and red) and about 55 per cent of plasma (watery solution of proteins and salts). Essentially the blood merely provides a medium to transfer materials to and from the main organs and tissues of the body. The red cells provide the transport for oxygen by means of a carrier called haemoglobin (which gives the cell a red colour when oxygen is present and dark blue when oxygen is absent). White blood cells exist in various forms (usually larger than the red cells), but their job is to engulf bacteria or produce antibodies which can react with foreign substances.

Problems with the blood system can arise as a result of an effect on the manufacturing site, i.e. the bone marrow, or an effect on the blood itself. Examples are:

(1) Carbon monoxide gas when inhaled passes into the bloodstream and reacts with the haemoglobin to form carboxyhaemoglobin. This is a stable material in that it does not easily revert back to haemoglobin. Since the haemoglobin is vital to carry the oxygen around the body, the loss of large quantities in this way will ultimately result in death. People suffering from carbon

monoxide poisoning therefore develop cyanosis which means that they take on a 'blue' pallor (particularly on the lips) indicating low levels of the 'red' oxygenated haemoglobin.

(2) Arsine is a gas that can be formed when a metal with an arsenic impurity contacts an acid and hydrogen is released. On absorption of arsine the chemical goes inside the red blood cells eventually rupturing the cells (haemolysis) and destroying them. The symptoms resulting therefore are related to the anaemia that the loss of red cells causes.

(3) Benzene is known to affect haemopoesis (the making of blood cells) and those suffering from *excessive* benzene exposure may therefore develop symptoms of anaemia, usually affecting the white cell count before the red cell count is seen to be lowered.

Target Organ 5: The Nervous System

The total nervous system is highly complex and will not be discussed in detail here. It is enough to say that some nerve systems involve the brain while others operate independently of the brain.

The 'messages' are passed along the nerves by chemical reactions stimulated by 'enzymes'. The nerves themselves are protected by a coating rather like the insulation on an electrical cable. This insulation is known as the *myelin sheath* and its existence is vital to the efficient working of the nerve.

Problems with the nervous system caused by occupational hazards are normally related to:

(1) interference with the enzymes that aid the conduction of information along the nerve;
(2) destruction of the myelin sheath protecting the nerve;
(3) reduction in the effectiveness of parts of the brain when certain chemicals penetrate the blood-brain barrier.

Examples of problem chemicals are:

(1) Some organo-phosphorus compounds (used as insecticides) can cause problems to the nervous system when excessively absorbed. These materials inhibit the action of the main enzyme involved in the conduction of nerve pulses. Usually the primary signs include ataxia (inability to stand upright with eyes closed

– loss of balance), but acute poisoning may lead to muscle twitching, coma and, ultimately, death.

(2) Mercury poisoning produces a symptom known as 'Hatter's Shakes' because it was seen amongst workmen dipping felt hats in an acid solution of mercuric nitrate (Mad Hatters!). This symptom usually begins in the fingers and then progresses to the arms and legs. A classic case of mercury poisoning appeared in the population around Minamata Bay, Japan. This was due to methyl mercury and caused slurred speech, numbness, unsteady gait and increasing disability affecting vision and hearing. Those seriously ill were mentally confused and severely agitated indicating a *'brain'* effect as well as an effect on the *peripheral* nervous system.

(3) N-hexane and methyl n-butyl ketone are recognised as potentially causing peripheral polyneuritis at high exposure levels. In both these cases the materials themselves are converted in the liver to the same metabolite which ultimately causes the problem. Peripheral polyneuritis means that the nerves at the extremities do not function adequately and hand tremor is usually a prime symptom.

Note that recently the use of equipment to measure nerve conduction velocities (how quickly the nerve responds to stimulus) and electromyography (whether the response of the nerve follows the same pattern as the 'input signal'), has meant that effects on the peripheral nervous system can be picked up *before* clinical symptoms appear.

Other topics of interest such as eyes, ears and skin are covered in specific areas in the text relating to their associated problems in industry. For further reading on these (and extended reading of the target organs discussed in this chapter) the reader is referred to textbooks of physiology such as *Basic Clinical Physiology* by J.H. Green, 3rd edn (Oxford University Press, Oxford, 1978).

Summary of Key Points

Lungs

'External' organ used to bring oxygen into contact with the blood and to remove waste gases such as carbon dioxide. Functional problems related to lung volumes and the condition of the alveoli.

Lung function tests include:

Spirometer	–	static lung volumes
Vitallograph	–	dynamic lung volumes
Pneumotachograph	–	small airways condition.

Liver

Chemical plant of the body:

Detoxifies protein breakdown products
Produces plasma protein
Produces heparin
Regulates glucose levels
Produces bile.

Metabolites formed in the liver from alien chemicals in order to solubilise for removal via urine. Excessive chemical exposure may divert liver from routine tasks and result in appropriate symptoms.

Kidneys

Prime function to produce urine to:

Maintain water balance
Maintain salt balance
Maintain blood pH.

About one million nephrons per kidney each having a glomerulus for contact with blood and to act as filter, plus a tubule to allow reabsorption back into blood. Main problems are with glomerulus and tubule failure, thus destroying the 'control'. Signs of failure are presence of blood, protein, glucose in urine or a change in the daily output rate of urine.

Blood

Red cells carry oxygen; white cells destroy bacteria. Manufactured in some of the bone marrow. Problems related to either the loss of blood production or direct effects on the cells themselves. Carbon monoxide and arsine affect red cell function whereas benzene affects their production.

Nervous System

Messages are passed along nerves by chemical changes induced by enzymes. Nerves have insulation called myelin sheath.

Problems associated with:

Inhibition of the enzymes
Destruction of the myelin sheath
Direct effects on the brain.

Examples of causative agents:

Organo-phosphorus insecticides
Methyl mercury
N-hexane
Methyl n-butyl ketone.

2 BASIC CHEMISTRY

Introduction

In industrial hygiene we deal with industrial hazards of many forms not the least of which is that due to 'chemicals'. Some of these are called 'organic' and some 'inorganic'. There can be few readers who have not heard of the terms organic and inorganic lead but unfortunately few people understand the differences between these compounds as a result of the use of the designations 'organic and inorganic'. In addition, to the uninitiated, references to materials such as hydrocarbon solvents, ketones and higher alcohols must strike fear in their hearts and bewilderment in their minds and once more they are forced to the familiar question – what does it mean?

If we expect you – the reader – to have a basic knowledge of industrial hygiene after reading this book, then one of the fundamental building bricks must be an understanding of basic chemistry. It is hoped that the next few pages will secure this for you.

What is Chemistry?

Chemistry is the science which deals with the properties of materials and their reactions with other substances.

Inorganic chemistry is concerned with the study of *non-living materials* such as minerals and metals and consequently, in industrial hygiene, we mainly associate inorganic chemistry with exposure to dusts, such as asbestos and silica, metal fumes as generated in welding and flame cutting, and mists produced during acid pickling or electroplating of metal.

Organic chemistry, as one would expect, deals with materials that are generally found in living organisms, both plant and animal. For example, petrochemical manufacture essentially reshapes and rebuilds *organic* chemicals which occur in very large quantities in nature (in living organisms). A great many medicines and pharmaceutical preparations are based on chemicals that are synthesised by plants, animals and – indeed – the human body.

Historical

Although use had been made of chemistry in early civilisations – prehistoric people were familiar with the production of wine by fermentation – a scientific approach to the subject started with the Greeks whose conception of the universe led them to postulate certain laws which they used as an aid to interpret scientific observation. This reasoning process was adopted by Plato, Aristotle and others and tended to overshadow true experimental reasoning in chemistry for hundreds of years – thus hindering its advance. Most of us are familiar with the early postulations of the Greeks that all matter consisted of the four elements of earth, air, fire and water and it was Aristotle who proposed that these could be further defined as hot, moist, dry and cold. It is remarkable that this theory continued until virtually the end of the eighteenth century and, indeed, we find that Aristotle's views had extended to medicine where it was believed that a healthy body had the four elements or humours of blood, phlegm, yellow bile and black bile in equilibrium. It was argued that disease resulted when this equilibrium was upset and could only be righted by the re-establishment of the body harmony. The almost cure-all remedy of blood-letting was, therefore, accepted on the basis of the equilibrium theory.

Paracelsus (1493-1541) founded a school which applied chemistry to the preparation of medicines and to the explanation of processes in the living body. Not only did Paracelsus revolutionise the scientific thinking of his day, he also applied careful 'study and observation' to the industries of his time and had published, in 1567, a study of the occupational diseases of mining and smelting workers. Paracelsus was not the first to mention diseases associated with mining, for some years earlier – in 1556 – *De Re Metallica* by Georgius Agricola (also known as Georg Bauer) was published in twelve volumes. This publication contained information on mining techniques, the ventilation of mine workings, on mining accidents and also details of lung diseases associated with this type of work. Bernardino Ramazzini (1633-1714) is regarded as the father of occupational medicine. His claim to that title results from his writings on occupational diseases which were published in his book *De Morbis Artificum Diatriba* in 1700. Ramazzini, in his time, noted the conditions of work and the occupational diseases of many trades as varied as miners, potters, cesspit cleaners and soap makers. In medicine at least we can see that by the early-eighteenth century there were men of ability who were unwilling to accept the status quo and were propelling scientific thinking to new and far-reaching frontiers.

Not only was medicine in a state of flux at this time but chemistry itself was advancing at a rate that had been unprecedented. Robert Boyle (1627-91) is regarded as the founder of modern chemistry for three main reasons. First, he regarded chemistry as worthy of study for its own sake and not merely as an aid to medicine and alchemy. Perhaps the main reason for his claim to the title of founder of modern chemistry is due to his introduction of rigorous experimental methods which allowed him to formulate laws from experimental data rather than following the early Greek technique of formulating laws as an aid to the explanation of events. Boyle also showed that the four elements of Aristotle and the three principle elements of the alchemists – namely mercury, sulphur and salt – did not deserve to be called true elements since none of them could be extracted (for example) from metals. J.J. Becher, a contemporary of Robert Boyle, however, still held to the earth theory and suggested that matter consisted of three types of earth – *terra pinguis, terra lapidia* and *terra mercurialis* (fatty earth, hard earth and softer earth respectively). He argued that combustible bodies lost the 'so-called' *terra pinguis* on combustion or calcination.

It should be noted that Boyle had carried out a number of combustion experiments but he had failed to explain how to interpret his results correctly.

A student of Becher's by the name of Stahl renamed this combustible *terra pinguis* as 'phlogiston'. This phlogiston was common to all materials and on calcination or combustion the phlogiston was given off. The residue left was called a calx and different substances had different calxes. Although lead gained weight when calcined, it was suggested that phlogiston had levitational properties and it buoyed up the metal which, therefore, gained weight when freed of phlogiston. Many chemical reactions could be explained in terms of this theory and hence it held sway for many years. The end to the theory of phlogiston came when Lavoisier (1743-94) carried out a series of rigidly controlled combustion experiments.

In the eighteenth and nineteenth centuries discoveries in chemistry continued apace and the majority of the basic laws were developed. Gases were studied, crystallography was developed, the combination of elements was beginning to be understood and later names like Faraday, Davy, Galvani and Volta were added to the long line of experimentalists.

Although in organic chemistry, processes for making soaps and the use of vegetable oils had been known for centuries, it was not until

Lavoisier's analytical technique was available that it was realised that compounds of organic origin contained mainly carbon and hydrogen. Between 1769 and 1785 Scheele isolated tartaric acid, citric acid, lactic acid and uric acid from grapes, lemons, sour milk and urine respectively. The development of organic chemistry was hindered due to the adherence to the vital-force theory or vitalism which stated that the synthesis of organic chemicals was impossible due to the absence of the essential life-force present in living systems. This theory was exploded by Wöhler in 1828 when he successfully produced urea (organic) from ammonium cyanate (inorganic). The structure of many compounds, including benzene, was known by the middle of the nineteenth century and many syntheses have been developed since Wöhler's early attempts.

Generally, organic compounds are combustible producing in the main carbon, carbon monoxide, carbon dioxide and water. As expected, organic materials tend to be insoluble in water but are usually soluble in other organic solvents. The majority of organic chemicals exist in nature either as gases, liquids or low-melting-point solids. In contrast the majority of inorganic compounds are solids, tend to be soluble in water and are generally non-combustible. These are general rules and exceptions may occur.

Definitions

An *element* is something which cannot be further divided into different chemical entities. Examples are metals, arsenic, carbon, phosphorus, hydrogen, oxygen, helium, etc. An *atom* is the smallest unit that is recognised as a particular element. All the atoms of one element are alike but they differ from the atoms of other elements. If we say that an element is like a heap of sand then the atoms – the smallest part recognised as an element – would be represented by the individual sand grains. A similar analogy can be used if instead of sand we had a pile of bricks – each individual brick would represent the atoms while all the atoms together produce the larger element, in this case the pile of bricks.

At the present time in excess of 100 elements are known. Some are very stable (e.g. lead), while others only exist in special conditions of temperature, pressure, etc. and this means that new elements continue to be isolated as research continues.

An atom consists of a number of parts and has a central core called a nucleus and an outer orbital shell. If we imagine the sun with the planets spinning around it, we get a simplified idea of the structure of the atom. The central core (the sun) contains positively-charged particles called protons and others called neutrons (which have no electrical charge). The electrons (planets) which spin around the nucleus (sun) are negatively charged and are equal in number to the protons so that the number of positive charges (protons) and number of negative charges (electrons) are equal – thus rendering the atom as a whole with a zero electrical charge. If we remove an electron (planet) from an atom it becomes positively charged (because we have one more proton than electrons) and if we add an electron the atom becomes negatively charged. Thus we form *ions* and if we add or remove more than one electron the ion has a net charge equal to the number of electrons we have added or removed, e.g.

add 1 electron, ion has 1 negative charge;
add 2 electrons, ion has 2 negative charges;
add 6 electrons, ion has 6 negative charges;
remove 1 electron, ion has 1 positive charge;
remove 2 electrons, ion has 2 positive charges;
remove 6 electrons, ion has 6 positive charges.

Positively-charged ions are known as *cations* and negatively-charged ions are known as *anions.*

In the lighter atoms, for example carbon, the nucleus consists of protons and neutrons in equal numbers. In the 'heavier' elements this is not necessarily true, for example lead has 82 protons and 125 neutrons in the nucleus or central core of the atom. If we add the numbers of protons and neutrons present we obtain the *atomic weight* of an element – carbon has an atomic weight of 12 while that of lead is 207.

Compounds are formed when atoms 'join together' in definite ratios and the smallest individual part of a compound that can be recognised as such is known as a *molecule.* In this respect molecules are to compounds as atoms are to elements. The following table may help in understanding this.

Elements	=	atoms plus atoms of same kind
Atom	=	smallest part of an element that can be recognised as that element
Compounds	=	atoms of one element plus atoms of another element (combining in definite ratios)
Molecule	=	smallest part of a compound that can be recognised as that compound

Molecules are held together by two main bonding forces. If we consider a molecule of sodium chloride or common salt, we see that it contains one atom of sodium and one atom of chlorine. The chemical symbols used to represent our sodium and chlorine atoms are Na and Cl respectively – so that our molecule of sodium chloride can be written symbolically as NaCl. If we remember our earlier discussion on the formation of ions and that some atoms can become more stable by losing an electron and become positively charged (cations), while others 'like to' gain electrons and become negatively charged (anions), we find when we look at sodium chloride the perfect combination in that sodium likes to lose an electron and chlorine likes to gain an electron (i.e. this is energetically more stable). We can represent these changes using the following chemical symbols:

Na minus an electron → Na^+
Cl plus an electron → Cl^-
$Na^+ + Cl^-$ → Na Cl

We can see that sodium and chlorine in the molecule of sodium chloride are held together by an electrical attraction formed between the positive sodium cation and the negative chloride anion. This type of bond is known as an ionic bond, i.e. a bond between ions.

A number of atoms on the other hand have no 'desire' to lose or gain an electron. If you like, they want to maintain their influence on all their electrons but at the same time they are hard pressed in their desire to achieve the most stable arrangement of electrons around their nucleii. Since either of these two distinct requirements cannot be totally met, a compromise position has to be accepted. Such atoms therefore share electrons with their neighbouring atoms. Carbon has four electrons and is able to share these in return for a share of another four electrons from its neighbours. By this sharing, carbon achieves the stable position of having a share in a total of eight electrons around its nucleus. Carbon can share with different atoms, for example hydrogen, or with neighbouring carbon atoms. Hydrogen has one electron and again, like carbon, is looking for a partner which will allow it to share another electron to achieve stability, that is when it has a share of two electrons. Hydrogen can, again like carbon, share with other different atoms or with a neighbouring hydrogen atom to form a hydrogen molecule. This sharing which binds atoms together is known as *covalent* bonding and can be represented thus:

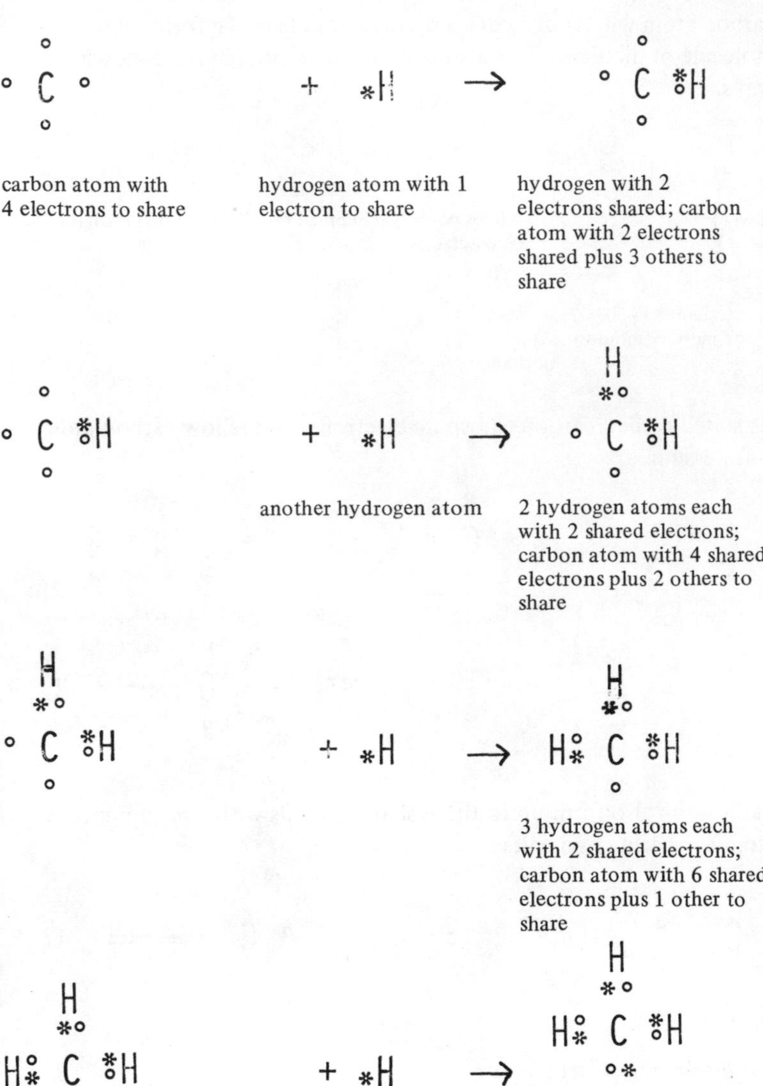

Carbon will always attempt to share eight electrons and thus will tend to combine with four other atoms. The combination of one carbon atom with four hydrogen atoms results in the formation of a molecule of methane. This also can be represented in the following ways.

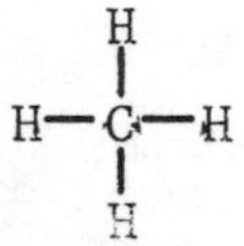

```
     H
     |
  H—C—H        where each – (chemical bond) represents a shared pair
     |          of electrons
     H
```

methane
or more commonly CH_4
methane

As stated earlier, carbon can share electrons with fellow carbon atoms – for example:

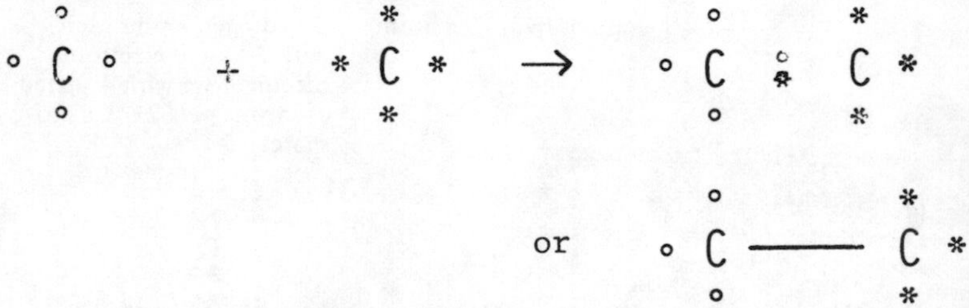

```
    o            *              o     *
  o C o    +   * C *    →     o C o*  C *
    o            *              o     *

                                o          *
                        or    o C ——————— C *
                                o          *
```

Carbon can also form up to three sharing bonds with a neighbouring atom by sharing two pairs

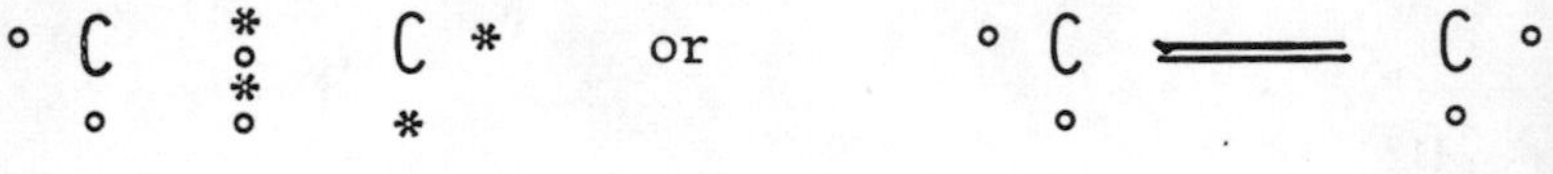

```
              *
  o C         o     C *      or      o C ═══════ C o
              *
    o         o     *                  o          o
```

and on sharing a further pair

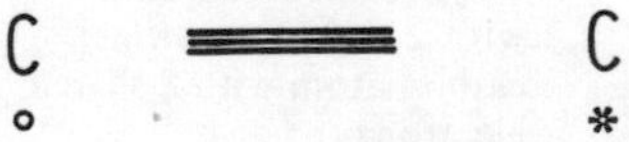

```
  C   ≡≡≡≡≡≡≡   C
  o             *
```

The remaining unshared electrons can be shared with two other carbon atoms, two hydrogen atoms or one of each:

$$C - C \equiv C - C \quad \text{or} \quad H - C \equiv C - H$$

$$\text{or} \quad H - C \equiv C - C$$

The above possible permutations and combinations may explain the vast number of organic chemicals that we find in nature. Each carbon atom strives to have a share in eight electrons while each hydrogen strives to share two. Carbon can also share electrons with oxygen, nitrogen and sulphur forming compounds called aldehydes, ketones, alcohols and acids with oxygen, amines and amides with nitrogen and sulphur compounds with sulphur. Shown below are a few examples of the simpler compounds found.

Structure	Formula	Name
H–C(H)(H)–H	CH_4	methane
H–C(H)(H)–C(H)(H)–H	C_2H_6	ethane
H–C(H)(H)–C(H)(H)–C(H)(H)–H	C_3H_8	propane
H–C(H)=C(H)–H	C_2H_4	ethylene
H–C(H)(H)–C(H)=C(H)–H	C_3H_6	propylene
H–C≡C–H	C_2H_2	acetylene
H–C(H)(H)–C(=O)–C(H)(H)–H	C_3H_6O	acetone

Structural formula	Formula	Name
$H-CH_2-C(=O)-CH_2-CH_2-H$	$CH_3CO\ C_2H_5$	methyl ethyl ketones
$H-CH_2-CH_2-OH$	C_2H_5OH	ethyl alcohol
$H-CH_2-CH=O$	C_2H_4O	acetaldehyde
$O=C(OH)-C(OH)=O$	$C_2H_2O_4$	oxalic acid
$H-CH_2-NH-H$	CH_3NH_2	methylamine
$H-CH_2-S-H$	CH_3SH	methyl mercaptan

Organic chemicals can be grouped into two main classes. Straight-chain compounds are known as *aliphatic* hydrocarbons and *aromatic* hydrocarbons possess an unsaturated (containing double or triple bonds) ring structure.

Aliphatic Hydrocarbons

Saturated chain compounds are called alkanes or paraffins. The term 'saturation' means that no double or triple bonds are present in the molecule (potential sharing with other atoms is used up). These compounds are relatively stable (non reactive), insoluble in water and, if heated to 400-600°C, will decompose or 'crack' to yield smaller-chain compounds or multiple-bond compounds, e.g.:

```
   H   H   H   H                      H   H            H  H
   |   |   |   |       heat           |   |            |  |
H--C---C---C---C--H   ----->   H--C---C--H   +    C≡C    +   H2
   |   |   |   |                  |   |
   H   H   H   H                  H   H

   butane                         ethane          acetylene   Hydrogen
```

Examples of alkanes are methane (CH_4); ethane (C_2H_6); propane (C_3H_8) and butane (C_4H_{10}).

In the alkane series we can replace one or more hydrogen atoms with atoms of chlorine to give us chlorinated compounds such as chloroform and carbon tetrachloride.

```
     H               H                 Cl
     |               |                 |
H----C----H    Cl----C----Cl     Cl----C----Cl
     |               |                 |
     H               Cl                Cl

  methane        chloroform      carbon tetrachloride

   CH4            CH Cl3              C Cl4
```

Alkenes or olefins are also chain-type compounds but they are distinguished from the alkanes by the presence of at least one double bond. In naming these compounds the ending -ane in the alkane series is replaced by -ene – hence the generic name of alkenes.

alkanes	*corresponding alkene*
butane	butene
C_4H_{10}	C_4H_8

```
    H   H   H   H              H   H   H   H
    |   |   |   |              |   |   |   |
H — C — C — C — C — H      H — C — C = C — C — H
    |   |   |   |              |           |
    H   H   H   H              H           H
```

but-2-ene

or

```
    H   H   H   H
    |   |   |   |
H — C = C — C — C — H
            |   |
            H   H
```

but-1-ene

The position of the double bond in these compounds adds an additional factor to the number of permutations and combinations in their structural layout.

It is possible for saturated aliphatic compounds to have a ring-type structure. This is seen with chemicals such as cyclohexane which, although a ring compound, is still aliphatic.

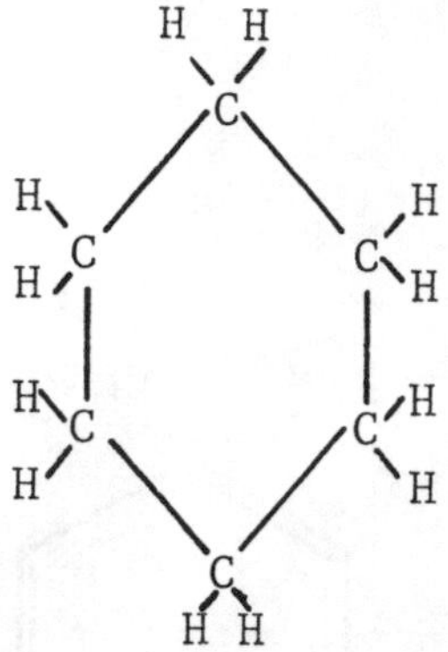

or

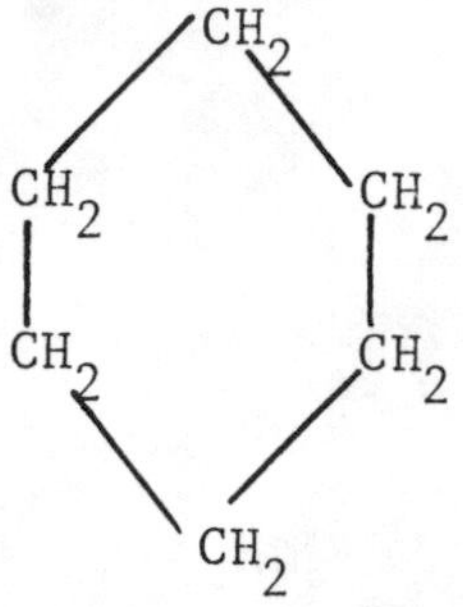

cyclohexane

Aromatic Hydrocarbons

Aromatic compounds are characterised by an unsaturated ring structure and the simplest in the group is benzene which possesses three double bonds.

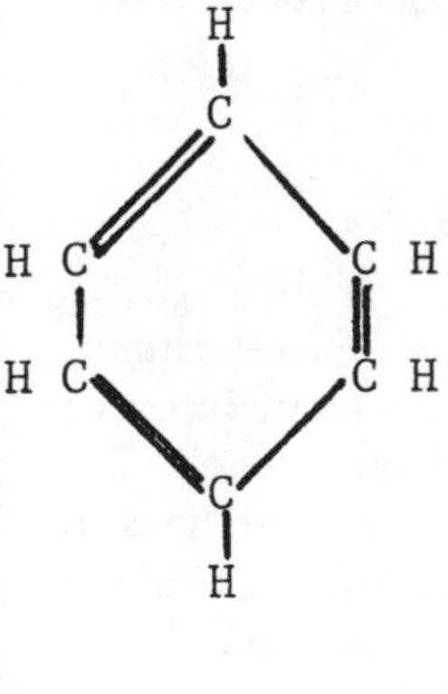

or

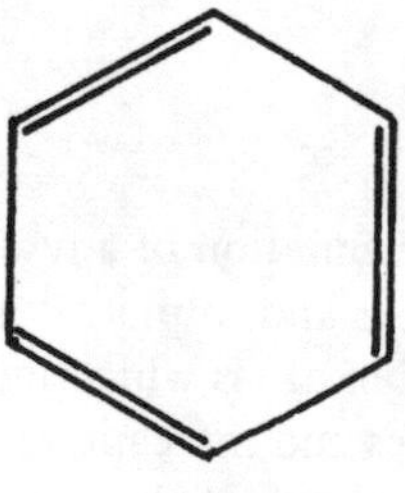

or

C_6H_6

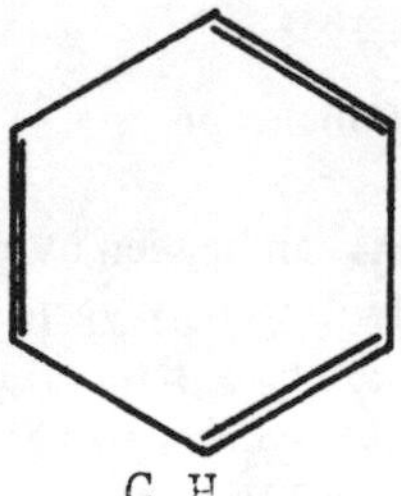

C_6H_6

By building on this structure we can produce many different compounds and the examples on page 36 show some of the simpler materials based on this basic building unit.

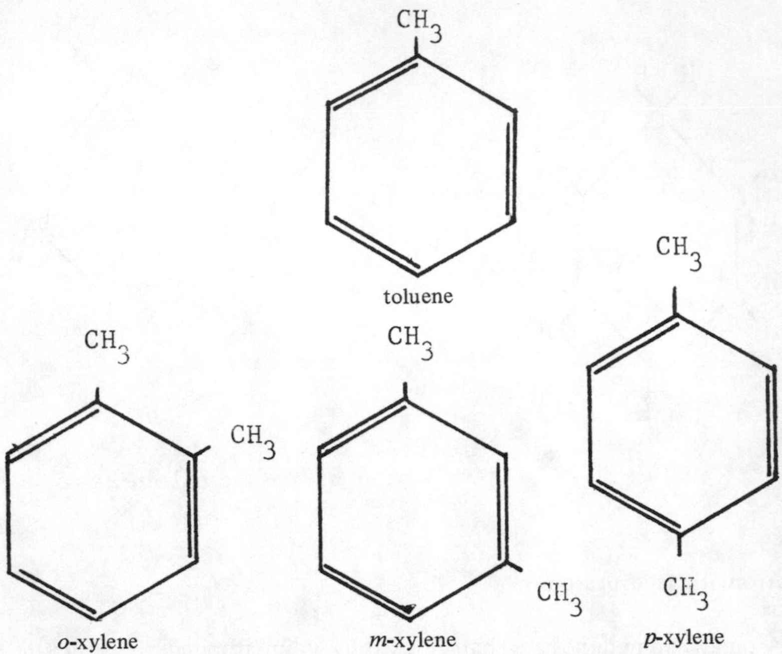

By introducing two side chains into the benzene ring we can produce three different variations on the same basic formula. The addition of two $-CH_3$ or methyl groups produces three xylenes – namely ortho (o), meta (m) and para (p) xylenes.

Conclusion

As can be seen by the combination of a few elements such as carbon, hydrogen, oxygen, nitrogen and sulphur it is possible to produce a wide variety of structural combinations which include acids, alcohols, aldehydes, ketones, amines and mercaptans. Further permutations are possible when the double and triple bonds are introduced and an even greater number of structures are possible using benzene-ring and straight-chain combinations. It should be noted that multi-ring compounds are also found so that the variety of chemical species continues to increase at an alarming rate.

Now the reader will realise that in a few pages it is impossible to cover the subject of chemistry in any depth as whole textbooks – sometimes in several volumes – are devoted to the subject but, by now, at least the terms organic and inorganic will hopefully be better understood. Consult the further reading list for some such textbooks.

Summary of Key Points

(1) A brief historical introduction is given along with basic definitions of terms such as atoms, molecules, ionic and covalent bonding.
(2) The difference between inorganic and organic compounds is stressed and examples are given to illustrate the way in which organic compounds can increase in complexity as additional 'building blocks' are added.

Further Reading

Fieser, L.F. and Fieser, M. *Organic Chemistry* (Reinhold Publishing Corporation, New York, 1956)
Finar, I.L. *Organic Chemistry*, volume 1 (Longman, London, 1968)
Hurd, D.J. and Kipling, J.L. *The Origins and Growth of Physical Science*, 2 vols (Pelican, Harmondsworth, 1964)

3 DUSTS, GASES AND VAPOURS

Introduction

The respiratory system, discussed in Chapter 1, is designed to allow the maximum blood-oxygen interface in the lungs and to allow the relatively free passage of air. This means that atmospheric contaminants readily gain access to the body and it is not surprising that in the working environment inhalation provides by far the highest risk of entry. The eventual fate of these materials will, in the case of dusts, be governed by their size and shape, and in the case of gases and vapours, by their solubility in body fluids.

Health Effects

In describing the toxic effects of dusts, gases and vapours, we must distinguish between short-term or immediate effects and those which may only be apparent after several years. The former are *acute* effects and result usually from a single exposure while the latter are described as *chronic* and usually result from repeated exposures over a number of months or years, generally at relatively low concentrations.

Health effects from 'mineral' dusts tend to be of the chronic type such as silica producing silicosis and asbestos, asbestosis. However, the irritant effect resulting from exposure to alkaline dusts would be considered an acute effect. Gases and vapours, on the other hand, are primarily associated with short-term acute effects and single exposures can produce unconsciousness, permanent damage to the lungs (as in the case of those exposed to phosgene, chlorine or mustard gas during the war years) or even death. There are obvious exceptions and repeated exposures at concentrations too low to produce any immediate effects may result in, for example: liver and kidney damage as in the case of chlorinated solvents such as carbon tetrachloride; leukaemia, as in the case of repeated excess exposures to benzene; or carcinogenic effects on the liver as in the case of vinyl chloride monomer.

Dusts

The general term 'dust' is used to describe airborne solid particles which vary in size and shape. The following terms are frequently used to describe airborne materials produced in specific processes and having a recognised size range:

Aerosol. A general name which refers to liquid or solid particles in air such as those generated by an 'aerosol can'.

Mists and Fogs. Liquid aerosol in air. Mists contain smaller droplets than fogs.

Dusts. Solid particles in air produced by sanding, discing or some other mechanical or abrasive operation.

Fumes; Solid particles produced by the volatilisation followed by condensation of metals as in welding and flame cutting. Fume particles are much smaller in size than dusts.

Smokes. Are similar in size to fumes and usually arise from combustion processes (or incomplete combustion of organic materials).

When considering solid particles, we need only distinguish between particles which are spherical (e.g. aerosols, mists and fogs, most dusts, fumes and smokes) and those which are fibrous, such as asbestos, cotton and flax dust, and glass wool. This distinction is important when deciding upon suitable sampling techniques.

Body Deposition

Spherical Particles. Dust sizes are usually expressed in micrometres (1 mm = 1000 micrometres, symbol μ). A human hair is of the order of 80-90 μ in diameter and the smallest particle likely to be seen with the naked eye is around 40 μ in diameter.

Dust particle sizes are most easily classified in terms of their aerodynamic diameter which takes into account the particle shape and density. For example, particles which have non-spherical shapes may have smaller aerodynamic diameters than spherical particles of the same density. It can be seen that the term 'aerodynamic diameter' is used to compare the behaviour of different particles in air because of their wide variation in shape and density. In this context the term diameter used to describe dust sizes refers to the aerodynamic diameter.

Particles with a diameter in excess of 10 μ are generally filtered out by the nose and trachea. Nose deposits are got rid of by sneezing or by blowing the nose. Cilia action moves tracheal deposits upwards and they are eventually swallowed. These particles although unable to reach the lungs can still result in irritation of the nose and trachea as in the case of acid or alkaline dusts.

Particles reaching the alveoli are called respirable dusts and have a size range of about 0.2-7 μ diameter. Dusts below 0.2 μ tend to remain in the airstream and are exhaled while those of about 7-10 μ are deposited in the upper respiratory system and removed by cilia action.

The sizes given above are general ranges only and do not have discrete cut-offs.

Fibres. The respiratory system is less efficient at trapping long, thin fibres and for this reason, those well in excess of 5 μ and even greater than 100 μ in length, as in the case of asbestos, may well reach the alveoli. In the case of fibres, their concentration (number per unit volume of air) is more important than their actual weight.

Threshold Limit Values

Threshold Limit Values (TLVs) are published annually by the American Conference of Governmental Industrial Hygienists (ACGIH) in the USA. These values are reprinted by the Health and Safety Executive and published in their Guidance Note series (EH15/79 is the latest edition at this time). TLVs refer to airborne concentrations of materials to which the majority of people may be repeatedly exposed without adverse effects. Acceptable exposure levels are set for working for eight hours per day and five days per week. It should be noted that TLVs should not be used to assess the difference between safe and unsafe conditions nor to assess the relative toxicities of materials. The aim should be to minimise all exposures where possible. In certain cases allowance is made for exposures in excess of the given TLV provided this is compensated for by periods below the level – so that the net average over the work period is at, or below, the TLV. These are called time-weighted averages and contrast with so-called ceiling-value TLVs which should not be exceeded at any time during the working day. Care must be taken in using TLVs because limits are set for varying reasons. In the case of ammonia the TLV is set to prevent undue irritation, while that for benzene is set to prevent the onset of long-term blood disorders.

Threshold Limit Values are not legally defined values but represent levels for good working practice which would be enforceable under section 2 (1) of the Health and Safety at Work Act of 1974 which states:

It shall be the duty of every employer to ensure, so far as is reasonably practicable, the health, safety and welfare at work of all his employees

Health Effects of Dusts

In terms of health effects, dusts can be classified into two main groups: those which may be regarded as inert or nuisance materials because their health risk tends to cause annoyance and inconvenience only, and those that produce a significant health risk.

Inert Materials. A TLV of 5 milligrams of dust per cubic metre of air (5 mg/m^3) is applied to the so-called respirable fraction of inert dusts in order to prevent a plugging of the respiratory system. Materials such as limestone, silicon carbide, Portland cement and gypsum are included in this group. Because of this they can be used as substitutes for other health-risk materials and it is not unusual to find limestone as a talc substitute for the dusting of bitumen products such as roofing felt to prevent adhesion in manufacture.

Siderosis – a benign pneumoconiosis – appears on the lung X-rays of those who have been exposed to iron oxide fume. These lung shadows are easily identified but do not appear to present any significant health risk.

In keeping with the principles of industrial hygiene, exposures to even so-called inert materials should be minimised where possible.

Dusts Producing a Health Effect. Dusts in this group may produce acute and/or chronic effects and also, in the case of allergic dusts, may produce an elevated disproportionate body response.

(1) Acute Effects. Alkaline mists and dusts such as slaked lime will produce nose and throat irritation which result in excess mucous production. The severity of such effects will depend upon both the concentration of the material in the breathing air, and the exposure time of the individual. Such materials will also produce eye and skin irritation.

(2) Acute and Chronic Effects. The continuance of excess mucous production may result in chronic bronchitis. Continued exposure to irritant materials would, therefore, be capable of producing both acute and chronic effects.

(3) Chronic Effects. Repeated exposures, usually measured in years, to a number of materials will produce a fibrosis of the lungs resulting in breathlessness and, in severe cases, premature death. The so-called pneumoconiosis dusts include crystalline silica which is often called free

silica, coal dust, talc and asbestos. Exposures to these produce silicosis, coal dust pneumoconiosis, talcosis and asbestosis:

(a) *Silicosis* is produced by the crystalline or free silica only and not by the amorphous (non-crystalline) material. Free silica exists in several forms including quartz, tridymite and cristobalite. The majority of materials dug from the ground will be likely to contain some free silica and its presence should be measured by specialist techniques such as X-ray diffraction as the dust TLV will decrease as the percentage of free silica increases. This should be checked when handling inert materials as the allowable exposure could decrease markedly if free silica is present, especially in its cristobalite or tridymite forms. Crystalline silica is found in sand (for mouldings in foundries), refractory bricks and cements, in the pottery industry in the clays used and in some filter aid materials. Older traditional industries such as slate quarrying have long been associated with 'dust disease' produced by crystalline silica.

(b) *Coal-dust Pneumoconiosis.* Coal dust probably contains free crystalline silica but the presence of other materials in the dust appears to result in the production of this recognised disease.

(c) *Talc.* It should be noted that talc may contain crystalline silica and in some cases asbestos, which would obviously enhance any lung damage resulting from talc exposure.

(d) *Asbestos.* As well as producing asbestosis, fibres of asbestos may also result in the onset of lung cancer and pleural mesothelioma which in the past has been associated with blue or crocidolite asbestos. Although exposures to the other forms – namely, white (chrysotile) and brown (amosite) – have not been clearly associated with this particular effect, they are under increasing suspicion. Smokers who are exposed to asbestos have a greatly increased risk as far as lung cancer is concerned.

(4) Allergic Effects. At low, and sometimes extremely low, concentrations, certain materials may give rise to an allergic response in a number of sensitised individuals. Hay-fever and sun-sensitive sufferers are not uncommon while in the furniture industry wood dusts – particularly those imported from so-called exotic areas of the world – are known to produce allergic responses. Exposure to vegetable textile dusts produces respiratory disease commonly known as byssinosis which may be caused by an allergen in the dust sensitising the bronchial

mucuous membrane. This disease is caused by exposure to cotton and flax dust and has been recognised for over 100 years. In their Hygiene Standards for Cotton Dust* the British Occupational Hygiene Society Committee on Hygiene Standards gives the following description of the disease:

> The characteristic symptoms of byssinosis occur on the first day back at work after a break. There may be cough, chest tightness or difficulty in breathing. The affected worker may first notice symptoms after annual holidays but later on they usually occur after weekends. Early effects may be noticed during the first year of exposure to dust and at this stage the first and only complaint may be cough or chest tightness after the work shift immediately following the weekend break (Mondays in Western countries). The cough, the feeling of chest tightness, or difficulty in breathing may disappear shortly after leaving the workplace. On Tuesdays there are no symptoms. As the disease progresses, the symptoms worsen and are accompanied by breathlessness. They extend to Tuesdays and then to other days, although at this stage of the disease there is still improvement as the week goes on. Eventually, the worker may become severely affected on every working day with chronic cough and sputum and permanent breathlessness which does not materially diminish, even on leaving the cotton industry. At this stage, the effects of cotton dust cannot be distinguished from chronic bronchitis, except that the past history of chest symptoms, characteristically worse at the beginning of the week, may suggest the aetiology.

Man-made Mineral Fibres (MMMF). Glass and mineral wools and ceramic fibres are included in this group of man-made mineral fibres. These materials are used as asbestos substitutes to an increasing extent and concern regarding their safety has been expressed. The literature to date seems to suggest that the chronic health effects associated with asbestos fibres are not seen with exposures to MMMF. However, the Health and Safety Executive has proposed exposure limits for these materials in its discussion document 'Man-made Mineral Fibres' (Report

**Annals of Occupational Hygiene,* vol. 15, no. 2-4 (1972). A BOHS standard for flax dust has recently been published by the Committee (*Ann. Occ. Hyg.,* vol. 23, no. 1 (1980)). These standards also detail monitoring techniques and give information on medical surveillance programmes.

of a Working Party to the Advisory Committee on Toxic Substances, available from HMSO).

Gases and Vapours

At ambient temperatures, gases and vapours behave similarly as far as their distribution in air is concerned. A gas is defined as a material which is in the gaseous phase under ambient conditions. Normally vapours result from the evaporation of materials which are liquid at ambient temperatures and pressures. The tendency for liquids to evaporate is governed by their vapour pressures. Those with higher vapour pressures will tend to evaporate to a greater extent and give rise to a higher atmospheric concentration than materials with a low vapour pressure under the same conditions. In general, as the liquid and the ambient temperatures increase the degree of evaporation increases.

Some solids also possess an appreciable vapour pressure at ambient temperatures and will, therefore, give rise to an appreciable vapour concentration in air. It is important to note that mercury – although liquid at ordinary temperatures – is a metal capable of being inhaled because of its ability to evaporate at ordinary temperatures.

If gases and vapours are insoluble in the body tissues they will tend to be exhaled, but if soluble they will tend to accumulate in the body until a saturation point is reached with an equilibrium existing between the atmospheric concentration and the amount absorbed. When exposure ceases the material will usually be exhaled gradually (or excreted by other routes) until none remains.

Gases and vapours can be classified into a number of groups according to their physiological effects as listed below.

Asphyxiants

The air we breathe contains approximately 20 per cent oxygen and 80 per cent nitrogen. This is an oversimplification of the truth but the other ingredients present in air are not relevant to this context. As the concentration of oxygen decreases in the air that we breathe, then the body becomes increasingly deprived of oxygen and a number of physiological effects occur as shown in Table 3.1.

Simple asphyxiants have, in themselves, no effects on the body system. Their asphyxiating power is due to the fact that their presence means that the concentration of oxygen decreases. For example, as normal air contains 20 per cent oxygen and 80 per cent nitrogen, then if we add 20 per cent of gas X the mixture would now contain 20 per cent of gas X along with only 16 per cent oxygen and 65 per cent nitrogen resulting in an increase in respiration and pulse rate.

Table 3.1: Physiological Effect of Changes in Oxygen Levels in Air

Oxygen percentage in air*	Physiological effect
20	None – normal concentration in air
18	Difficulty concentrating
16	Emotional instability, confusion, drunk feeling
12–16	Increased respiration and pulse rate
10–12	Lack of muscular co-ordination, fatigue, slight cyanosis
6–10	Inability to move, nausea and vomiting, loss of consciousness
5	Gasping for breath, convulsions, respiration failure and death. Minimal concentration compatible with life
4	Loss of consciousness and convulsions in 40 seconds
2	Immediate loss of consciousness and convulsions and respiratory failure
0	Immediate coma. Brain damage in 4-5 minutes

*after Gerarde, H.W., private communication.

The so-called inert gases such as helium, neon and argon and materials such as nitrogen, methane and carbon dioxide are all simple asphyxiants (although some data are available which indicate that carbon dioxide may have other chronic effects).

Chemical asphyxiants on the other hand interfere with the body's oxidative processes in some way, thereby causing significant health risk and death – even in situations where the oxygen concentration is still able to maintain life. Thus carbon monoxide combines with haemoglobin forming carboxyhaemoglobin. Aniline liquid is readily absorbed through the intact skin and vaporises at room temperatures and can, therefore, be readily inhaled. Once absorbed aniline combines to form methaemoglobin and thus – like carbon monoxide – reduces the oxygen uptake by the haemoglobin. Hydrogen sulphide which has a distinct rotten-egg smell acts upon the central nervous system causing respiratory paralysis. This material is extremely fast-acting and a concentration as low as 700 ppm can be rapidly fatal. Cyanides prevent the uptake of oxygen by the body cells and may produce death even when the blood is fully saturated with oxygen.

Irritants

Irritants are classified according to their effects on the respiratory system and it is fair to say that their atmospheric concentration is more important from a response point of view than the exposure time. Irritants such as formaldehyde, hydrogen chloride, hydrogen fluoride and sulphur dioxide tend to be relatively easily soluble in mucous and therefore readily exert their irritant effect on the upper part of the respiratory system. It should, however, be noted that some material will pass into the lower respiratory system and in sufficient concentration will produce a chemical pneumonitis.

Due to their slightly lower solubilities, materials such as bromine, chlorine, fluorine and iodine tend to be classified as both upper- and lower-respiratory irritants. Ozone, which is formed by irradiation of atmospheric oxygen by ultra-violet light, as found in welding, can produce severe damage to the lungs and is certainly not to be sought out as some beneficial elixir, as the Victorians previously assumed.

In the 1914-18 war, gas warfare included phosgene in its arsenal and those exposed suffered from pulmonary oedema. This material can also be produced along with hydrochloric acid as a result of the decomposition of chlorinated solvents (such as trichloroethylene) by ultra-violet light generated during welding. In flame cutting and welding, nitrogen dioxide – another lower-respiratory irritant – is formed by the combination of atmospheric oxygen and nitrogen at the high temperatures encountered in the processes.

Anaesthetics and Narcotics

Virtually all solvent vapours have a depressive effect on the central nervous system. Their ability to produce such effects varies from substance to substance but in general at concentrations of several hundred parts-per-million these effects will be marked. In addition to their narcotic effect a number of solvents can cause injury to specific organs of the body which are not themselves in direct contact with the solvent. These are known as systemic effects, while the susceptible organ of the body is known, not illogically, as the target organ. Such organs may be affected by either the solvent itself or by some metabolite manufactured in the body's biochemical factory. Fat-soluble materials will tend to deposit in adipose tissue while the liver and kidneys – because they are both involved in the metabolism and excretion of materials – are common target organs for a large number of materials, including chlorinated solvents. The activity of the enzyme cholinesterase is inhibited by organo-phosphorus pesticides and the blood-forming system is subject to attack by benzene which may produce leukaemia if repeated exposures are sufficiently high. Generally, the concentrations producing anaesthetic effects are well in excess of the levels of repeated exposures regarded as capable of producing these chronic systemic effects.

Sampling for Dusts, Gases and Vapours

In industrial hygiene we are primarily interested in the exposure that a worker receives during day-to-day work. For this reason samplers for dusts, gases and vapours have been developed that can be worn by the operators as they go about their duties. Hence such equipment should be small, lightweight and should not present an additional burden to the operator. In addition, we are concerned about both shift and shorter-term or peak exposures so that the monitoring technique chosen must be capable of giving reliable and accurate results when used for various periods. If full-shift or part-shift samples are collected, they can be used to assess the degree of risk based on a time-weighted basis. 'Instantaneous' sampling is used to measure peak concentrations or can be used to assess exposures during some event in the work process such as the collecting of plant samples for quality-control purposes. Sampling periods must also take account of the detection limits of the method used and we must ensure that a sufficient collection time is chosen to allow these criteria to be met.

Area samples, as opposed to personal samples, can provide information about the concentration of a contaminant in an area and can be used to assess, for example, whether entry into the area is permissible. They can also be used to assess overall background concentrations and can, therefore, be useful in indicating when a plant upset occurs. This information can be used to provide an evacuation warning to employees and highly sophisticated, multi-sample point, automatic systems which have developed for this purpose.

Dusts

Small personal sampling pumps (see Figure 3.1) collecting between 1 and 4 litres per minute (2-litres-per-minute pumps are commonly used) are connected via tubing to a filter holder sited near the operator's

Figure 3.1: Typical Dust Sampling Pumps. Left: high volume site (static) sampler. Middle: 1-4 litres/minute pump (used for personal samples). Right: 1-4 litres/minute pump (also used for personal samples but with automatic flow control).

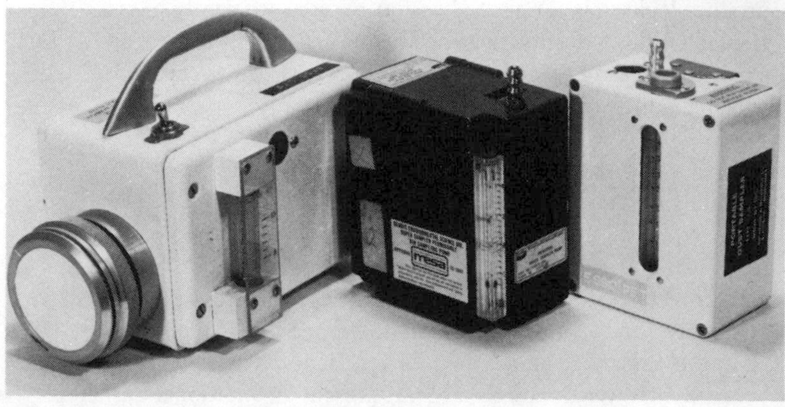

breathing zone. In practice the filter is usually fixed to the lapel. If respirable dust is collected, then a sample head capable of the separation of dust according to size is required and the most common type in use is the cyclone, developed by the British Cast Iron Research Association (see Figure 3.2). If no size selection is required, then an open-faced filter holder is generally used. Asbestos fibres and metal fumes are collected in this way. The open filter should be held vertically and not horizontally so as to avoid picking up the 'fall-out' of large particles not normally inhaled.

Filter papers made of a variety of materials are available and care must be taken to ensure that the correct choice is made for a particular application. If attempts are made to measure the dust concentration gravimetrically, then the filter paper used needs to be relatively unaffected by moisture. In addition, a balance capable of measuring to 0.01 mg should be available – particularly if low dust concentrations are expected.

Subsequent chemical analysis may also govern the choice of paper. Asbestos fibres are collected on membrane filters which are chosen so that they can be made transparent when mounted on a microscope slide. This is necessary to allow the fibres to be counted using phase-contrast microscopy.

Metal fumes are usually collected on papers and then analysed chemically and again the filter media should allow the material to be

Figure 3.2: Sampling Heads. Used for respirable dust (cyclone in the middle of the picture) and total dust (on right of picture). The actual cyclone in the middle of the photograph fits inside the protective cover shown on the left.

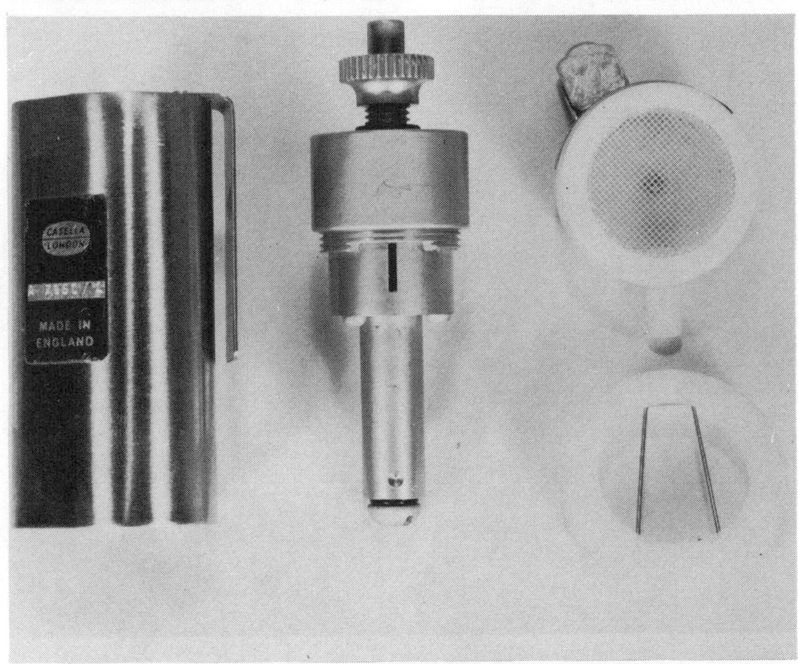

transferred easily into solution for this purpose.

In the main, dust and fume concentrations are measured in milligrams of dust per cubic metre of air collected (mg/m^3) and for both the respirable and total size ranges. Asbestos fibres on the other hand are measured in the number of asbestos fibres per cubic centimetre of air collected ($fibres/cm^3$ or fibres/ml) while MMMF can be assessed by both methods.

Gases and Vapours

Both gases and vapours can be collected using personal sampling pumps and an adsorbent filter medium worn in the breathing zone (see Figures 3.3 and 3.4). Pumps used in this case have much lower flow rates than those used for dusts and usually sample between 5 and 100 cm^3/min.

The choice of adsorbent is again governed by the material under investigation and the subsequent analytical techniques which may

Figure 3.3: Low-flow Personal Sampling Pump Used in Connection with Adsorbent Tube for Gases/Vapours. Photograph indicates back and front to show 'clip-on' facility for belt, pocket, etc. of wearer. The sample head shown is the NIOSH charcoal tube.

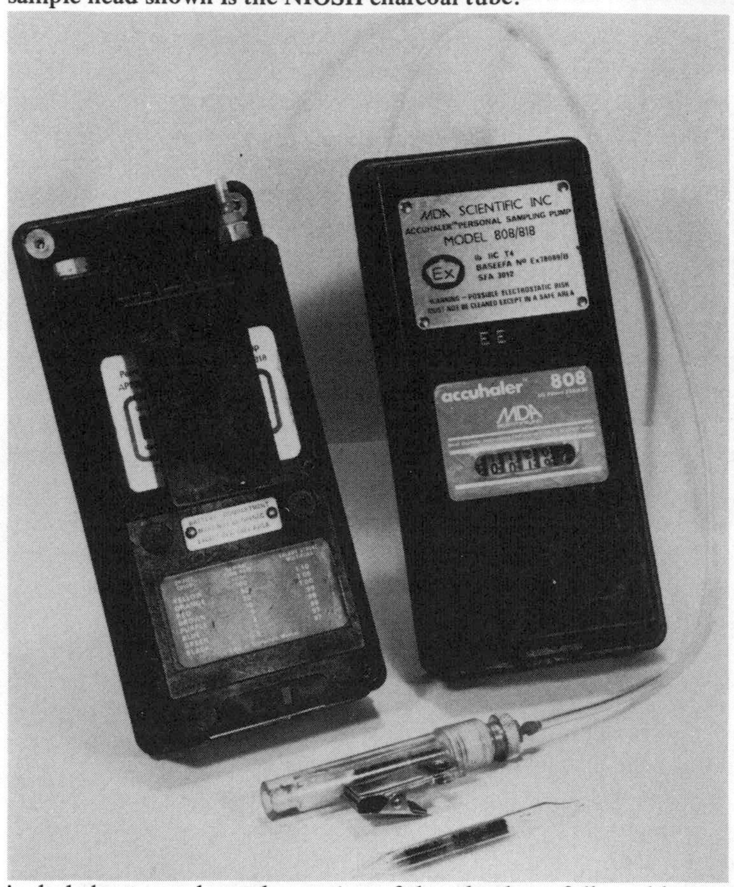

include heat or solvent desorption of the adsorbent followed by accurate analysis by gas chromatography.

Detector tubes can be used for assessing both peak or short-term, and shift exposures. Different tubes for individual gases and vapours have been developed and contain crystals of particular chemicals which give a colour change when exposed to their specified gas or vapour. Generally a pre-determined volume of air is collected and the stain length or intensity is calibrated to give a concentration in parts of gas or vapour per million parts of air (ppm by volume usually). In the main, these tubes are used with a hand-held pump and the sample is collected

Figure 3.4: Self-compensating Low-flow Pump. Sample head is an EMS heat-desorber type of collecting system. (Supplied by Bastock Marketing as in list of suppliers at end of this book.)

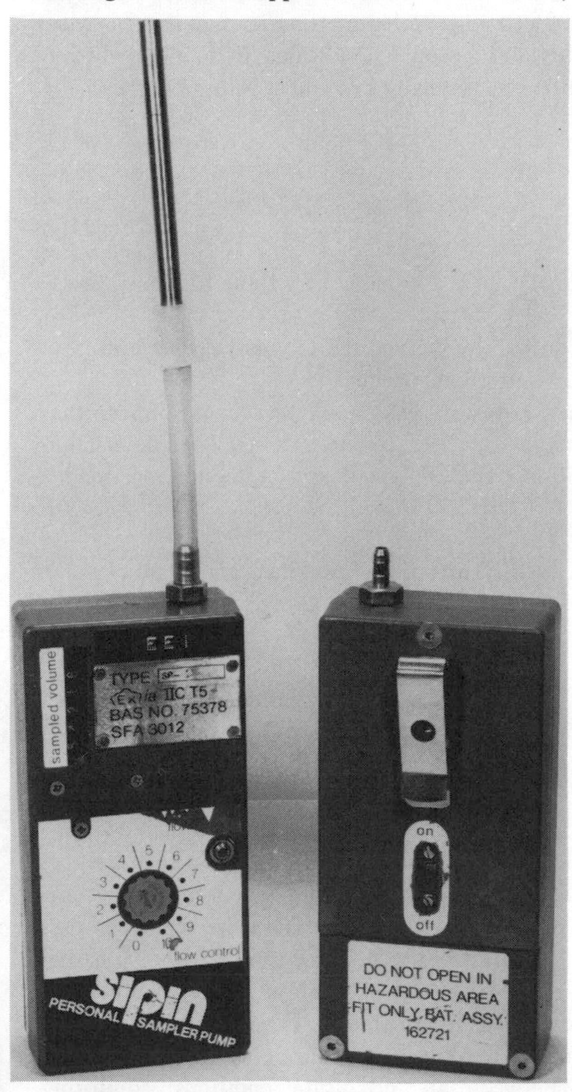

over a few minutes. By holding the detector tube near the breathing zone one can assess the operator's exposure over a short period, for example during a specific task. Special long-term tubes for use with low-flow pumps have been developed for sampling over several hours so that personal exposures can be assessed.

When using sampling pumps, care must be taken to ensure that they will not present a fire risk when used in potentially flammable atmospheres such as found in petroleum refineries, and in areas where volatile solvents are used under open conditions. Equipment which has been approved (certified intrinsically safe) must be used in these situations.

Summary of Key Points

(1) In industry, inhalation is the main entry route for dusts, gases and vapours.

(2) Control procedures may be required to guard against both acute and chronic exposures and effects.

(3) In assessing dust exposure, one must take account of both the size of the particles and their concentrations. When determining exposures to fibrous materials such as asbestos and man-made mineral fibres both the length and the number of airborne fibres must be taken into account.

(4) Dusts can be classified into those producing an instant effect on mucous membranes and the respiratory system and those producing chronic effects on the lungs – pneumoconiosis and lung cancer – or systemic toxicity.

(5) Gases may be divided into two main groups, simple or chemical asphyxiants. Those in the latter category have a definite physiological effect, while those in the former group merely produce an effect because the per-cent oxygen concentration in the breathing air has been decreased to an unacceptable level.

(6) Vapours in relatively high concentrations have an anaesthetic and/or narcotic effect. Some, for example benzene and carbon tetrachloride, produce chronic effects following repeated exposures at concentrations well below those producing narcotic effects.

(7) To assess exposures personal sampling is recommended.

(8) Threshold Limit Values are for guidance only and should not be used to decide between safe and unsafe conditions. Monitoring results should be assessed by someone trained in the use of TLVs and as a general principle exposures should be kept as low as are reasonably practicable.

Further Reading

NIOSH *The Industrial Environment – its Evaluation and Control* (US Government Printing Office, Washington, DC, 1973)

Patty, F.A. *Industrial Hygiene and Toxicology,* 3 vols (John Wiley, New York, 1978)

4 METALS

Introduction

Metals were regarded as useful materials by early man just in the same way as other items found in his environment. The development of metallurgy as we know it has grown markedly since these early times and even now new metal alloys and metal uses are continually in development. Like virtually all substances in our environment, the use of metals brings untold benefit while their misuse brings risks.

It is hoped in the next few pages to give some information on the potential health risks associated with a few of the metals in common use today. In no way will an attempt be made to present a complete exposition on the subject and the reader is directed for this to other reference works as listed at the end of the chapter.

General Health Effects

As mentioned in Chapter 3 metals in the form of dust, fumes and, in the case of mercury, as vapour, may get into the body. The prime entry route is via inhalation, although poor standards of personal hygiene will mean that material in significant quantities can be ingested. Organo-metal compounds such as tetraethyl and tetramethyl lead which are used as anti-knock additives in gasoline, can be absorbed through the intact skin and this has to be guarded against by the use of impervious clothing and the prompt treatment of skin contamination.

Metal dusts are produced during the mining, crushing, drilling and the general working of metal ores, metals themselves and their alloys. Welding and flame cutting operations result in fume production. Although dusts may be trapped by the nose and upper respiratory system, they will still gain entry to the body system if they are soluble in mucous fluid. Certain metal oxides, for example vanadium pentoxide, can be formed during combustion of some oils, and these and chromic acid mist, used for chrome plating, are respiratory irritants. Nickel can produce both local skin effects and can also act as a skin sensitiser producing skin reactions, sometimes remote from the point of body contact. Certain chromates are regarded as carcinogenic while lead and mercury can produce systemic effects in certain body organs (in this sense the nervous system is regarded as a 'target' organ).

In order to be helpful, general data on some of the commoner metals in use today are summarised below. The list is not intended to be exhaustive and the reader is directed to the reference material listed for additional information. Where Threshold Limit Values (TLVs) are quoted they are taken from Guidance Note EH15/79 published by the Health and Safety Executive.

Lead 

Care should be taken to distinguish between inorganic and organic lead compounds (see Chapter 2).

Inorganic Lead Compounds

Uses. These compounds have been used:

(1) in the past for water-pipes and cisterns;
(2) as a flashing material in roof construction;
(3) in the manufacture of paints;
(4) as a lead glaze in the pottery industry;
(5) in the manufacture of batteries;
(6) as a lining material for steel containers for sulphuric acid use;
(7) in the manufacture of alloys;
(8) as a solder.

Uptake. Inorganic lead is taken up as a dust generated by grinding, drilling and working of lead and lead-painted or alloyed steels and the handling of lead compounds. The spraying of lead paint is now prohibited. Lead fume is generated during welding and flame cutting of lead-painted or lead-alloyed steels and during the smelting, casting and moulding of lead. Poor standards of personal hygiene may result in the ingestion of significant amounts of material. This is a particular problem in the case of young children as a result of 'pica' (the eating and chewing of objects in this case contaminated with lead as e.g. lead paint and lead-contaminated dusts). Lead can also be ingested from food and water. In the normal diet this is unlikely to be significant but the use of lead water-pipes and cisterns can result in significant concentrations in drinking supplies – particularly in soft-water areas. It has been known for significant uptake to occur in the home brewing of wine due to the use of lead-glazed storage vessels from which the lead is dissolved by organic acids.

Health Effects. The more severe or more striking health effects, including peripheral neuritis (lead palsy or wrist drop), the lead or blue line in the gums, tremor and encephalopathy, are extremely rare in industry at this time. Colic and anaemia are still more common but their incidence is probably on the decrease due to the enforcement of improved preventive measures. The initial symptoms of lead poisoning include sleeplessness, a general fatigue or malaise and constipation. Such relatively vague symptoms may not specifically indicate lead intoxication, but biological monitoring including the determination of the levels of lead in blood and urine can assist in the clinical diagnosis by confirming absorption.

It should be noted also that Section 82 of the Factories Act 1961 requires that a medical practitioner must inform the Health and Safety Executive of the names of any patient suffering, or whom he believes is suffering, from lead poisoning.

Legislation. The latest legislation covering employment in lead processes is The Control of Lead at Work Regulations, made 1980 and coming into operation on 18 August 1981 (SI 1980 no. 1248). A new Code of Practice for the Control of Lead at Work has just been issued by the Health and Safety Executive to coincide with the introduction of these regulations.

There is a requirement under the regulations to assess exposures using both atmospheric and biological monitoring. Where these are significant (in excess of half the TLV) routine monitoring must be instituted. In addition, adequate medical examinations must be undertaken and all records kept for a minimum of two years. The use of protective clothing and exhaust ventilation and facilities for washing and changing, eating, drinking and smoking are also covered by the regulations. The present Threshold Limit Value for inorganic lead fumes and dusts is 0.15 mg lead per cubic metre of air (0.15 mg/m^3).

Organic Lead

Uses. Tetraethyl and tetramethyl lead (TEL and TML) are called organic lead compounds due to the combination of lead with organic molecules (see Chapter 2). These materials are used as anti-knock additives for gasoline and are generally added to the gasoline as part of the refinery process.

Uptake. TEL and TML are oily liquids at ambient temperatures and they possess a sufficiently high vapour pressure (give off vapours from the liquid easily) to result in appreciable vapour concentrations in the air. Therefore, their main uptake route is by vapour inhalation – although significant amounts can pass through the intact skin if this occurs. TEL and TML tend to accumulate in the sludge and scale of gasoline storage tanks and several cases of poisoning and even deaths have occurred from the cleaning of these vessels by inadequately protected workers. The handling of gasoline containing low concentrations of blended TEL/TML does not present a significant health risk from lead, providing the TLV for gasoline vapour is not exceeded.

Health Effects. The severity of the symptoms varies with the duration of the exposure and the concentration of the TEL and/or TML in the air. Insomnia, malaise, loss of appetite and gastro-intestinal disturbances are regarded as symptoms of mild intoxication, while mental confusion, excitement, sleeplessness, nausea and vomiting indicate more severe poisoning. In fatal cases, death may be preceded by acute mental disturbance, delirium, mania, convulsions, hallucinations and coma.

It should be noted that urinary lead concentrations are a somewhat better indicator of TEL and TML uptake than blood lead levels. Although TEL and TML are present in gasoline, they are largely emitted in car exhausts as inorganic compounds due to their conversion in the combustion process. Cases of organic lead poisoning would also have to be notified under Section 82 of the Factories Act 1961.

Legislation. The general regulations for inorganic materials will also apply to the handling of organic lead compounds. In addition, the Associated Octel Company of Ellesmere Port issues recommended procedures on a wide variety of aspects of organic lead handling, varying from the handling of TEL and TML through to tank cleaning and spill clean-up. The present Threshold Limit Value for TEL is 0.1 mg/m^3 and for TML is 0.15 mg/m^3.

Mercury

As with lead, mercury (chemical symbol Hg) is commonly used in both inorganic and organic forms and we must distinguish between the two when considering their particular health effects.

Inorganic Mercury

Uses. Metallic mercury uses include the following:

(1) in electrical switches;
(2) in thermometers and barometers;
(3) in the manufacture of amalgams, including those for dentistry;
(4) as an electrode material in electrolysis as in chlorine manufacture;
(5) in the manufacture of mercury compounds.

Inorganic mercury compounds are used in the manufacture of:

(1) pigments and anti-fouling paints for ships (mercuric oxide);
(2) pharmaceuticals (mercurous and mercuric chloride);
(3) timber preservative (mercuric chloride);
(4) for 'carrotting' in the felt-hat industry (mercuric nitrate).

Uptake. Metallic mercury and some of its compounds produce an appreciable vapour concentration at ambient temperatures resulting in an inhalation risk. The dust of inorganic compounds can also be inhaled and uptake through the intact skin for both metallic mercury and the inorganic mercury compounds is possible.

Health Effects. In industry, acute poisoning is rare. Chronic poisoning is more likely – although its appearance has decreased markedly over the years. Early symptoms may include nausea, malaise, headaches, stomatitis and diarrhoea. As mentioned earlier, mercuric nitrate is used for carrotting felt in the hat industry and one marked chronic effect of mercury poisoning is psychological disturbance or mercurial erethism resulting in shyness, loss of confidence, coupled with vague fears and depression. In advanced cases there may be a loss of memory and hallucinations may also be found. These changes resulted in the common phrase 'as mad as a hatter' because of their prevalence in that industry. Stomatitis, muscular tremors, bleeding gums, excessive salivation and a metal taste in the mouth are all indicative of chronic poisoning.

Assessment of the levels of mercury in urine can give an indication of excessive absorption.

Organic Mercury

Uses. Organic mercury compounds are used in the manufacture of:

(1) seed dressings designed to inhibit fungal growth and delay germination (phenyl, ethyl and methyl compounds);
(2) bulb dips (methoxyethyl mercury chloride);
(3) materials to control the growth of slime in paper mills (phenyl mercuric acetate);
(4) detonators for explosives (mercury fulminate).

Uptake. In industry, uptake is primarily by inhalation of the dust or liquid aerosol during compound handling. On the other hand, significant cases of poisoning have occurred in the general population – most notably in Minamata Bay in Japan caused by mercury compounds present in factory effluent. Although the effluent was in the inorganic form, bacteria in the bay converted it to methyl mercury which was absorbed by the fish and thence to the community via the food chain. In another case, seeds treated with an organic compound (dressed) for planting were ground to flour for baking and were responsible for mass poisonings in Iraq in 1956 and 1960.

Health Effects. Methyl and ethyl mercury compounds (so-called alkyl compounds; see Chapter 2) tend to act on the central nervous system producing tiredness and numbness and tingling in the fingers and toes. As the severity increases, tremors, loss of co-ordination of movement (ataxia) and difficulty in speaking may occur. Total physical disability is not unknown. Aryl compounds such as phenyl mercury acetate tend to produce health effects similar to those produced by the inorganic compounds.

Skin Effects. Some inorganic and organic compounds can have marked local effects on the skin. For example, mercuric chloride which is soluble can denature the proteins present in the skin. Blistering of the skin is caused by contact with phenyl mercury acetate and vesicular dermatitis is produced by mercury fulminate.

Legislation. Mercury poisoning is a notifiable disease under Section 82 of the Factories Act of 1961. The present Threshold Limit Values are: mercury (all forms except alkyl compounds), 0.05 mg/m^3; alkyl compounds (on the skin), 0.01 mg/m^3.

Manganese

Uses. Manganese is used in the manufacture of:

(1) alloys including those with steel;
(2) in dry-cell battery manufacture;
(3) in chemicals manufacture;
(4) and as a flux in welding rods and wires.

Uptake. Uptake is generally as a result of the inhalation of manganese dusts generated during material handling or by fume inhalation during welding and flame cutting.

Health Effects. Acute effects appear to be limited to pneumonitis – a condition similar to influenza – and is also produced by the inhalation of other heavy metal oxides including zinc and copper (brass founders' ague). Chronic manganese poisoning is associated with psychological and neurological disorders. Excessive salivation, blood changes and pulmonary symptoms have all been reported. Psychological and neurological disorders include headache and pain in the joints and muscles. Hallucinations and mental confusion are not uncommon. Due to delayed muscle response a mask-like face is a characteristic of chronic poisoning with additional symptoms frequently including tremors and Parkinson-disease-type effects.

Legislation. Manganese poisoning – both acute and chronic – are notifiable under an amendment of the Factory and Workshop (Notification of Diseases) Order of 1936 and incorporated in Section 82 of the Factories Act 1961. The Threshold Limit Value for manganese dust is 5 mg/m^3. This is a so-called ceiling value 'C' notation and should not be exceeded at any time. For manganese fume the TLV is 1 mg/m^3.

Chromium

Uses. Chromium metal is widely used to produce alloys with:

(1) iron – stainless steels;
(2) nickel – nichrome resistance wires for electric fires, etc.;
(3) vanadium – for hard steels;

(4) cobalt and tungsten – surgical instruments.

Chromium compounds are used for:

(1) chromium plating;
(2) paint and pigments (e.g. lead chromate);
(3) tanning leather (e.g. sodium dichromate);
(4) timber preservation (e.g. chromated zinc chloride);
(5) dyestuffs.

Uptake. The inhalation of dusts and liquid droplets from plating baths, etc. is the prime uptake route.

Health Effects. In the main, hexavalent chromium (in chromates, dichromates and chromic acid) is of prime concern as far as potential health risks are concerned. Due to their corrosive action, chromic acid and its salts are irritating to the skin and mucous membranes of the respiratory system. Ulceration of the skin and nasal septum is not uncommon and perforation of the latter frequently occurs where exposure is excessive. New workers particularly may find local skin irritation a problem. This usually disappears when exposure continues. In susceptible cases chronic eczema may result. Asthmatic effects may also result where high exposures to chromic acid occur. Lung cancer due to chromate and dichromate production from chromite-ore has also been reported.

Legislation. The occurrence of chrome ulceration is notifiable under Section 82 of the Factories Act 1961. The Chromium Plating Regulations 1931 (as amended by statutory instrument No. 9 in 1973) requires that either exhaust ventilation be provided at chrome-plating baths to prevent the emission of spray or vapour into the work area or that the baths be covered or treated in some way so as to prevent this. The testing of these precautionary measures including atmospheric sampling has to be carried out every 14 days. Additional requirements of the regulations relating to washing of floors and personal protection are also required.

Additional legislation includes the following:

(1) Chemical Works Regulations 1922
(2) Chrome Ulceration Order 1919
(3) Dyeing (use of Bichromate of Potassium or Sodium) Welfare

Order 1918
(4) Tanning (Two Bath Process) Welfare Order 1916

The TLVs for various chromium compounds (as metal concentration) are: chromium metal (as chromium), 0.5 mg/m^3; chromium divalent and trivalent compounds (as chromium), 0.5 mg/m^3; chromium hexavalent compounds (as chromium), 0.05 mg/m^3.

Cadmium

Uses. Cadmium is used:

(1) as a protective coating for iron, steel and copper (electroplating, dipping and spraying);
(2) in the manufacture of low-melting-point alloys suitable in alarm systems and automatic sprinkler systems;
(3) as a constituent of some paints and printing inks;
(4) in silver solders.

Uptake. The main route for intake in industry is by inhalation of the dust or fume produced by grinding, drilling, etc. of cadmium-plated and alloyed materials and by inhalation of the oxide produced during the heating of cadmium-containing materials during, for example, welding, flame cutting and silver soldering. Cadmium may be present in foodstuffs but only some 5 per cent of the ingested material is absorbed via the gastro-intestinal tract.

Health Effects: Acute. Single exposures to fume can result in an acute inflammatory reaction of the respiratory tract which may prove fatal. The severity of the reaction increases with increasing fume concentration. It should be noted that the onset of symptoms including malaise, fever, tightness and pain in the chest, an irritating cough and dyspnoea, may be delayed for anything up to 24 hours. For this reason persons suspected of over-exposure should be kept under observation for at least 24 hours.

Health Effects: Chronic. Chronic effects of exposure can affect both the renal and respiratory systems. Excessive absorption of cadmium can result in the presence of low-molecular-weight proteins in the urine which may persist over a period of years. Again, the onset of symptoms may be delayed up to 15-20 years after initial exposure. The presence of these proteins in urine is not, however, specific to cadmium exposure. Respiratory impairment can be detected by lung function tests including the forced expiratory volume in one second (FEV_1), the forced vital capacity (FVC) and the vital capacity (VC). Respiratory impairment is progressive and again a latent interval between initial exposure and the onset of symptoms is likely. Those suffering from respiratory insufficiency may show radiological changes associated with pulmonary emphysema.

Legislation. Poisoning by cadmium or its compounds is notifiable under Section 82 of the Factories Act 1961. The present Threshold Limit Values applicable to cadmium are: cadmium dust and salts (as cadmium), 0.05 mg/m^3; cadmium oxide fume (as cadmium), 0.05 mg/m^3; cadmium oxide production (as cadmium), 0.05 mg/m^3 (suspect carcinogen).

The British Occupational Hygiene Society in their 'Hygiene Standard for Cadmium' published in *Annals of Occupational Hygiene,* vol. 20 (1977), pp. 215-28, recommend a hygiene standard of 0.05 mg/m^3 for respirable material and of 0.20 mg/m^3 for that portion of total dust sample soluble in 0.1N hydrochloric acid. A Special Short Exposure Limit (SSEL) of 2 mg cadmium/m^3 *for a maximum of ten minutes* is acceptable, provided no further exposure occurs for the rest of the shift.

Cobalt

Uses. Cobalt is mainly used as an alloy constituent with chrome, nickel, aluminium, copper and molybdenum. The addition of cobalt to steels improves the cutting ability of tools manufactured from them and cobalt is widely used in the manufacture of tungsten carbide tools.

Uptake. In industry, uptake is mainly by the inhalation of cobalt-containing dust and fume.

Health Effects. Gastric disturbances, allergic dermatitis similar to that produced by contact with nickel, and lung and respiratory effects have all been reported.

Legislation. The present TLV for cobalt dust and fume (as cobalt) is 0.1 mg/m^3.

Nickel

Uses. Nickel is widely used in the manufacture of alloys with metals including iron, chromium, manganese and copper. So called 'monel' alloys are primarily nickel-copper and contain about 70 per cent nickel. Permanent magnets contain nickel, aluminium, iron, cobalt and other elements. Corrosion-resistant steels also contain nickel. Nickel anodes and nickel salts are widely used in the electroplating industry, while nickel compounds are used as catalysts in refining and chemical processes.

Uptake. Dust and fume inhalation represents the prime route for body entry. These, however, do not appear to produce significant health effects in normal industrial use (however, see below). In the Mond production process, nickel carbonyl is formed by combination of nickel and carbon monoxide. The carbonyl is subsequently decomposed to give pure nickel. As the carbonyl is volatile, it can be inhaled if it escapes into the work area.

Health Effects. The main cause for concern when handling nickel, its alloys or its salts, is its ability to produce an allergic dermatitis which not only may arise at the site of body contact but also can result in skin effects on other parts of the body. Once this sensitivity has developed it is virtually permanent.

Lung cancer has been reported in Norway where a nickel refinery was using an electrolytic process for the production of nickel.

Nickel carbonyl exposure produces respiratory distress including chest pain, dry cough, shortness of breath, rapid respiration, cyanosis and extreme weakness. Lung and nasal cancer have also been associated with exposure to nickel carbonyl.

Legislation. The present Threshold Limit Values are for nickel metal 1 mg/m^3 and for nickel carbonyl (as nickel) 0.05 ppm. Soluble nickel compounds (as nickel) have a TLV of 0.1 mg/m^3 while in sulphide roasting the TLV for nickel fume and dust is 1 mg/m^3 (carcinogen). Occupational cancer due to exposure to nickel powder 'formed by decomposition of a gaseous nickel compound' has been a prescribed disease since 1949, although confusion as to the exact carcinogen still exists.

Zinc

Uses. Zinc is used:

(1) as a protective coating to prevent corrosion in iron and steel, being applied by galvanizing, spraying or dipping;
(2) as zinc sheet in building;
(3) in the manufacture of brass which contains up to 40 per cent zinc;
(4) as an alloy constituent with copper, nickel, aluminium and magnesium;
(5) as an alternative to cadmium in electroplating;
(6) in paints and in paper.

Uptake. The main route for absorption is via inhalation of the dust or fume and also the oxide. Ingestion of zinc compounds can result in gastro-intestinal symptoms but in the main this route of absorption is not significant.

Health Effects. The inhalation of zinc oxide fume results in 'metal fume fever' or 'brass founders' ague' which resembles influenza in its effects. The condition usually lasts for 24 hours and recovery is complete. Workers develop an immunity to the disease but this is readily lost.

Deaths resulting from exposures to high concentrations of zinc chloride fume have been reported.

Chronic disease resulting from exposures to zinc fume and dust appears to be unlikely due to the presence of other additional materials which have been implicated. However, regular episodes of the acute effects may tend to lower the body's resistance to other diseases.

Legislation. The Threshold Limit Value for zinc chloride fume is 1 mg/m^3 and for zinc oxide fume 5 mg/m^3. A TLV of 0.05 mg/m^3 has been set for zinc chromate (as chromium), a suspect carcinogen. Zinc oxide dust is regarded as a nuisance particulate with a TLV of 5 mg/m^3 respirable dust.

Copper

Uses. Copper has many uses in industry and these include:

(1) the manufacture of a wide range of alloys including brass and bronze;
(2) due to its high thermal and electrical conductivity it is ideally suited for use in the electrical industry for resistance wires, electrical conductors, arc-air gouging rods, etc.;
(3) pipes for both gases and water;
(4) copper salts as insecticides in horticulture.

Uptake. Copper is a natural constituent of the human body and intake from food is around 2-5 mg per day. Ingestion, however, is not of major significance in industry where the main absorption route is by the inhalation of dust and oxide fume.

Health Effects. Various reports have suggested both acute and chronic health effects that were attributable to the inhalation of copper fume and dust from copper compounds. In the main it has been accepted that the only significant health effect is 'metal fume fever' caused by inhalation of the oxide. It should be noted that both the acetate and sulphate salts are acutely toxic even when ingested in small amounts.

Legislation. The present TLV for copper fume is 0.2 mg/m^3 and for dusts and mists (as copper) is 1 mg/m^3.

Beryllium

Uses. Beryllium is used mainly for:

(1) the manufacture of 2 per cent copper-beryllium alloys for non-sparking tools and in the electronics industries;
(2) the manufacture of X-ray tubes, as a phosphor for fluorescent lamps, etc. (The use of beryllium for lamps has now effectively ceased due to its toxicity and safer substitutes such as calcium halophosphate are now available.);
(3) nuclear reactors.

Uptake. The dust generated during the handling of beryllium compounds is inhaled and readily absorbed by the body and is the main route of uptake – sometimes through ignorance of the fact that beryllium may be present, for example, during the handling of broken fluorescent tubes.

Health Effects. Acute beryllium disease is a chemical pneumonitis characterised by a cough, burning chest pains, anorexia and fatigue, progressive dyspnoea and increased respiratory rate. In the main, recovery is complete but if exposure has been severe death may result. Acute dermatitis, ulceration and granulomas may also result from exposure to beryllium compounds.

Chronic beryllium disease is associated with increasing respiratory distress. Initial symptoms may include shortness of breath and cough progressing to chest pains, weight loss and general malaise as the disease increases in severity. Death is eventually due to cardiac or respiratory failure. Chronic skin lesions are associated with the disease and may continue to occur even after exposure has ceased.

Legislation. Poisoning by beryllium and its compounds is notifiable under Section 82 of the Factories Act 1961. Beryllium disease is also listed as a prescribed disease under the National Insurance (Industrial Injuries) Act 1946. The present TLV for beryllium is 0.002 mg/m^3 (suspect carcinogen).

Vanadium

Uses. Vanadium is used in:

(1) the steel industry to produce improved steel in terms of hardness and fatigue resistance;
(2) as a catalyst in chemical processes;
(3) in insecticides.

Uptake. In the main, uptake is via inhalation of the dust of vanadium compounds, primarily of the pentoxide. Vanadium in the form of metavanadates has been used therapeutically in the past.

Health Effects. Significant exposure to vanadium pentoxide dust may occur during the cleaning of oil-fired burners and combustion chambers. The renewal of the refractory brick can, in these circumstances, result in significant exposures.

Health effects are primarily associated with the respiratory system and include nasal catarrh, nose bleeding, chest pains, dry cough and wheezing and dyspnoea. Acute bronchitis may also result and nausea, vomiting and abdominal pain have also been reported. Chronic

disorders do not appear to predominate – although prolonged exposure has produced tremor in the hands. Palpitation on exertion has been frequently reported.

The most striking evidence of vanadium exposure is the discoloration of the tongue which turns greenish-black. This disappears when exposure ceases but it cannot be cleaned from the surface of the tongue.

Legislation. The Threshold Limit Value for vanadium pentoxide dust (as vanadium) is 0.5 mg/m^3 and that for the pentoxide fume is 0.05 mg/m^3. This latter TLV is a ceiling value.

Measurement

In the main, personal samplers fitted with suitable sample heads and filter papers can be used to assess operator exposure to metal dusts and fumes. The methods employed will be governed by the subsequent analytical techniques available in the laboratory. Fume samples should be collected using an open filter head while dusts can be sampled using the BCIRA size selective head which collects the respirable dust fraction. In both fume and dust sampling a flow rate of 2 l/minute is acceptable.

Lead alkyls are collected on an absorbent material such as Porapak Q, using a low-flow pump sampling at 100 cm^3/minute or less. This technique allows personal samples to be collected. Absorbed lead alkyl is then heat-desorbed into the flame of an atomic absorption instrument fitted with a lead-detector lamp. Area lead and lead alkyl levels can be determined using a bubbler and colorimetric techniques.

Mercury vapour can be detected using a direct reading instrument having an ultra-violet light source.

Conclusion

It is impossible to cover the metals above in great detail and the aim has been to give general information on a few of the more commonly used metals. Doubtless a number will disagree with the choice made. The reader is recommended to consult other authors – especially Ethel Browning and Donald Hunter – for more detailed and wider information. Indeed, in preparing this section, a great deal of the information presented has been drawn from these two sources.

Summary of Key Points

(1) Inhalation is the prime body entry route for metals which may be present as dusts (from grinding, crushing) or as fumes (welding, flame cutting or smelting).
(2) Exposures to metals may occur when working the metal, metal alloys or painted or galvanised surfaces.
(3) Organic and inorganic compounds of a particular metal may produce differing health effects, e.g. lead and mercury compounds.
(4) Monitoring should be assessed using personal sampling pumps.

Further Reading

Browning, E. *Toxicity of Industrial Metals* (Butterworths, London, 1961)
Department of Health and Social Security *Lead and Health* (HMSO, London, 1980)
Department of the Environment 'Lead in Drinking Water', pollution paper no. 12 (HMSO, London, 1977)
Health and Safety Executive *Cadmium* (EH/1), *Chromium* (EH/2), *Prevention of Industrial Lead Poisoning* (EH/3), *Chromic Acid Concentrations in Air* (EH/6), *Beryllium* (EH/13) (HMSO, London, published periodically)
—— 'Threshold Limit Values', Guidance Note EH15/79 (HMSO, London, 1979)
Hunter, D. *The Diseases of Occupations* (English Universities Press, London, 1975)
Lead Development Association *Medical Aspects of Lead Absorption in Industrial Processes* (Lead Development Association, London, 1973)

5 NOISE

Introduction

In terms of industrial production our modern, high-speed and sophisticated technological world may not be as successful as we would wish. This contrasts markedly with our ability to produce unwanted sound, more commonly called *noise.* In energy terms, wastage as noise represents a minute loss, but this is overshadowed by the monumental efforts that are made to reduce that loss so that we can live quieter lives.

Health Effects

Noise has three main effects on the exposed population. If we are subjected to high noise levels for a sufficient period of time then a deafness or hearing loss will result. At lower noise levels deafness may not occur, but difficulties in communication may result and we may find it impossible, for example, to use the telephone, as in the case of using a telephone kiosk with a few panes of glass missing near a busy road, or to hear warning signals such as fire and evacuation alarms. Noise may also produce annoyance and irritation, for example when we fail to get to sleep because of the neighbour's cat singing romantic overtures or when we cannot distinguish the gun-fire on the recording of the 1812 Overture from the rumbling and banging produced by the factory next door. It is relatively simple to issue guidelines and actual numbers which we can apply to prevent deafness and ease communication. Specifications are also available which allow us to set criteria which will prevent annoyance to the majority of the population but, as will be appreciated, annoyance criteria are by far the most difficult to apply as any parent will tell you when the latest number-one sound blares from the radio.

Hearing Damage

In 1972 the Department of Employment issued a 'Code of Practice for Reducing the Exposure of Employed Persons to Noise' which suggested that over a working lifetime of 30 years a deafness risk associated with a noise exposure of 90 decibels 'A' weighted (written 90 dBA) for eight hours per day and five days per week was acceptable. It should be noted that decibels are measured on a logarithmic unit scale with a sound level meter. Incorporation of a bias in the response of the sound level meter so that our results simulate the behaviour of the human ear in our measurement gives us our so-called 'A' weighting. Because our decibel scale is based upon logarithms, a doubling or halving of our sound intensity means that our measured level increases or decreases by 3 dB. If we have a machine producing 90 dBA and we introduce a second machine producing 90 dBA then the combined noise level would be 93 dBA and *not* 180 dBA. The 'Code of Practice' states that 90 dBA for eight hours per day is acceptable for non-protected persons. If we double the sound energy that we are exposed to, that is increase the noise level to 93 dBA, we should halve our daily exposure period to 4 hours. In summary, if we double the sound level that we are exposed to then we must halve the exposure time. This concept of a noise dose (a noise level for a set time) is known as the equal energy concept and the 'Code of Practice' uses this in determining the acceptable, or allowable, daily exposure periods if no hearing protection is worn. This is illustrated in Table 5.1.

Table 5.1: Allowable Exposure to Noise Levels

Noise level in dBA	Allowable daily exposure (hours per day, 5 days per week)
90	8
93	4
96	2
99	1
102	0.5
etc.	etc.

In order to grasp what the sound of a particular level is like, the noise levels associated with a number of activities are shown in Table 5.2.

Noise may be considered as a series of waves issuing from a source in a similar manner to the waves on a pond when a stone is thrown. These sound waves enter the outer ear and move to the ear drum (tympanic membrane). When the waves hit the ear drum it starts to vibrate and these vibrations are transmitted by a three-boned mechanical linkage to the 'oval window' in the inner ear. The names of these bones (the hammer, which is attached to the ear drum; anvil; and stirrup which is attached to the oval window) aptly described their action which serves to amplify the small movements of the ear drum. Behind the oval window, that is on the brain side, we have a snail-shell-like tube filled with fluid. Running down the centre of this tube there is a membrane covered with minute hair cells: As the oval window moves, waves are sent through the fluid in the tube and the hair cells respond to these movements and send messages to the brain telling us that a high noise level is being heard. When hearing damage occurs it is the hair cells in the inner ear that are affected and there is no technique available for restoring their use. Amplification as found with a hearing aid will not overcome the problem. Indeed, in noise-induced deafness the use of a hearing aid would be similar to increasing the volume on a poorly-tuned radio.

Table 5.2: Noise Levels of Everyday Activities

Sound level (dBA)	Activity
30-50	private office
50-65	general office
60-70	speech
70-90	average traffic
90	raised voice – shouting
>120	painful to the ears

Different hair cells in the inner ear respond to different frequencies of noise. The frequency of a sound is expressed in Hertz (Hz), formerly cycles-per-second, and is used to describe the pitch of the sound that we hear. Steam or compressed-air leaks are described as having a high frequency while the the noise produced by diesel engines would be predominantly of low frequency. The time-signal pips on the radio have a frequency of 1,000 Hertz while middle 'C' has a frequency of 256 Hertz. In the main, high frequency sounds tend to be more damaging than those of lower frequency. In order to assess the likely damage to hearing we must take account of the intensity of the noise in decibels, the frequency in Hertz and the exposure time in hours per day.

Noise damage usually begins with the hair cells unable to respond to frequencies around 4,000 Hertz (4 kilo Hertz, written 4 kHz) and as normal speech lies in the 500-4,000 Hz frequency range, the initial deafness will pass unnoticed. As the deafness progresses, both the frequency ranges affected and the extent of damage increase, so that by the time a person is aware of this it is too late for prevention. It should be noted that occupational deafness does not result in total silence for the sufferer, who finds difficulty in interpreting what is said rather than not hearing anything. As consonants are spoken at higher frequencies than vowel sounds, one can see that as hearing efficiency at these frequencies decreases then the ability to distinguish words must also diminish. Noise-induced hearing loss is added to the normal ageing losses (presbyacusis) that occur and the effects of an induced loss become more marked as age increases.

In order to prevent deafness occurring, we must determine the noise exposures of the operators and compare these with the recommendations given in the Code of Practice.

Measurement of Noise Exposures

As mentioned previously, noise dose involves an exposure to a given noise level for a set period of time. In practice, it is unusual for the plant or factory noise level to remain constant throughout a working day. We mentioned earlier the relationship between the actual noise energy received over a given period of time and we concluded that an exposure of 90 dBA over eight hours was the same as 93 dBA for four hours, 96 dBA for two hours, and so on. We can, using a sound-level meter (see Figure 5.1) determine the noise level in particular locations. If that level is fairly constant and the workforce remains in the area for most of the day, then our measured level will record the noise exposure of the individuals and we can decide whether that exposure is acceptable by referring to the Code of Practice. If, however, our noise level fluctuates randomly throughout the day or if our workforce moves between noisy and quieter areas within the factory, then it is more difficult to estimate exposures. This information can be obtained using a noise-average meter or personal noise dosimeters (worn by the operators) which take account of both quiet and noisy periods and 'average' the results giving a single figure on the output scale. This single figure is called an L-equivalent (written L eq) and it means, in effect, that our complicated exposure pattern of fluctuating noise levels and times spent in these levels result in a noise dose being received equivalent to x dBA for eight hours – x would be the L eq value obtained from our noise-average meter or noise dosimeter. If our exposure times are fairly constant, for example two hours at 90 dBA, followed by one hour at 100 dBA, followed by three hours at 92 dBA, followed by two hours at 87 dBA, then using our sound-level meter to determine these various levels and our knowledge of the work process

Figure 5.1: Precision-grade Sound-level Meter. Attachments are also available for octave band analysis, extended microphone arm, tripod and taping connections for the recording of sound for future laboratory analysis. Also shown are standard ear muffs and two types of disposable ear plugs, i.e. glass wool and expansion foam.

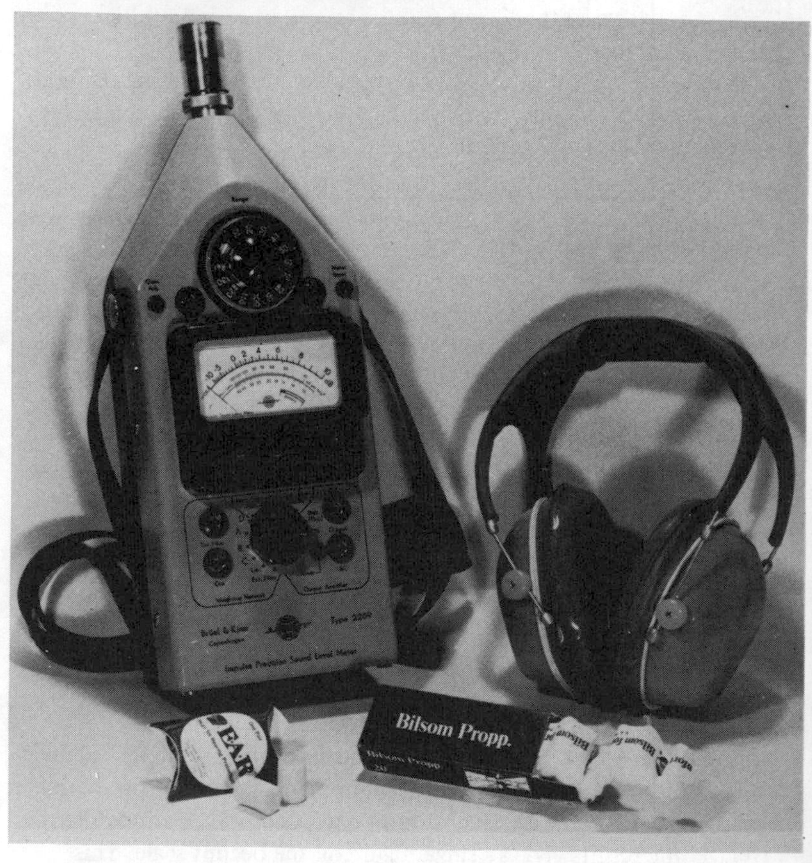

to determine the exposure times, we can calculate our L eq for this type of exposure pattern. Details on how to do this are given in Appendix 3 in the Code of Practice and in this case, our L eq would be approximately 94 dBA. In summary, varying work situations are shown diagrammatically in Figures 5.2, 5.3 and 5.4.

Figure 5.2: Steady Exposure – Use Sound-level Meter to Assess

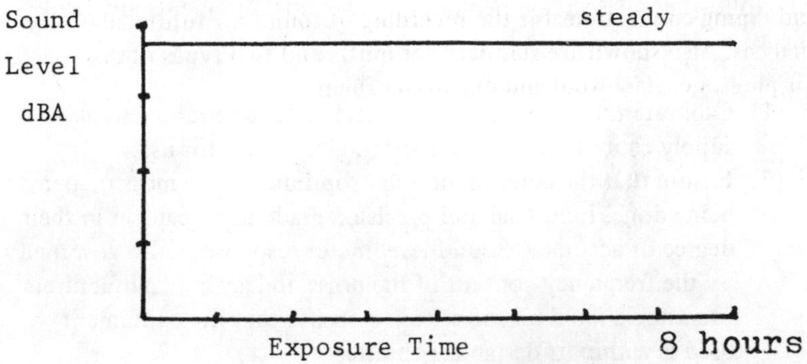

Figure 5.3: Worker Moves between Areas of Different but Constant Sound Levels. Exposure is assessed with sound-level meter to assess all noise levels, plus estimate/measure of exposure time at these levels.

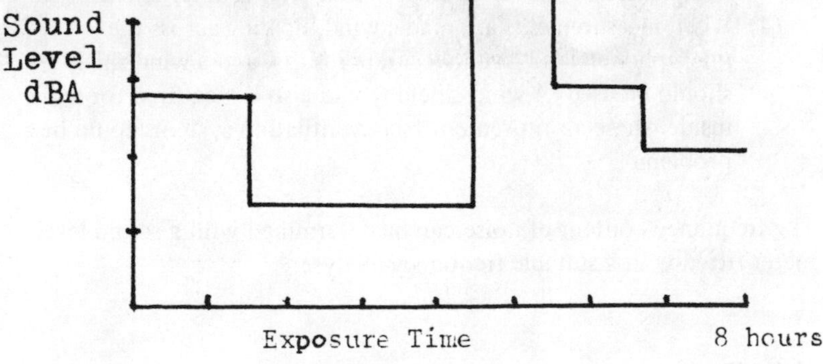

Figure 5.4: Sound Level Always Varying. Use noise-average meter or personal noise dosimeters.

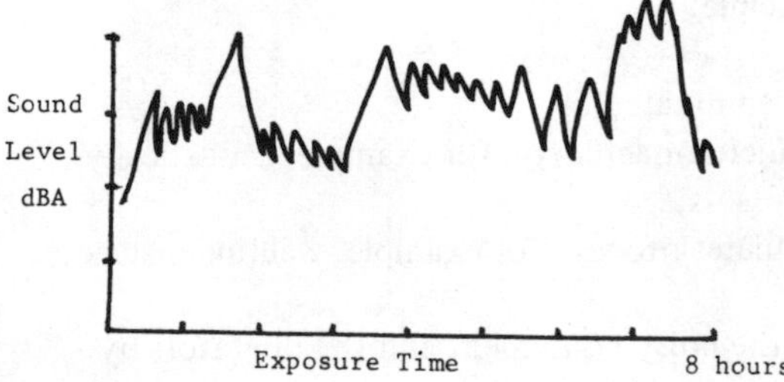

Measurement Technique

The following rules should be adopted when measurements are being made:

(1) Calibrate the equipment – manufacturers of equipment also supply calibration equipment and give details for use.
(2) Ensure that the equipment is appropriate for the measurements being done. Industrial and precision-grade meters differ in their degree of accuracy. Sound-level-meter response is also governed by the frequency content of the noise and again manufacturers' literature should be consulted to ensure that the equipment used is within its design capabilities.
(3) When assessing operator exposures the microphone of the sound-level meter (and noise-average meter) should be held at the operator's ear position in his or her normal work station. In other situations the meter should be held away from the body and panned so that an average sound level is determined.
(4) When measurements are made, wind blowing across the microphone could give inaccurate results, and a wind-shield should be used. A wind-shield would also be required for work inside where air movement from ventilation systems could be a problem.

The frequency content of noise can be determined with a sound-level meter fitted with a suitable frequency analyser.

Noise Control

To control noise effectively, a stepwise approach should be followed and by answering the following questions the basic control procedures can be developed in principle:

(1) Can the noise be eliminated by:
    (a) the use of quieter machinery, for example, acoustically quiet compressors;
    (b) by using a quieter process, for example, welding instead of riveting?
(2) Can we separate the noisy equipment and the operators by:
    (a) enclosing the noisy equipment or

(b) by enclosing the operators?
In both (a) and (b), machinery controls can be manipulated in a quiet area.

(3) *Only* if engineering controls are not possible either because of engineering difficulties or because of cost, then hearing protection may be considered. It should be noted that the initial cost of protection may be far below any engineering control, but in time personal protection may be more expensive for the following reasons:
(a) equipment replacement due to wear and tear and loss;
(b) education and training of the workforce in its correct use particularly for new recruits;
(c) maintenance, cleaning, etc. of the equipment.

Enclosures

The design of enclosures should be left to an expert as a knowledge of the frequency content of the noise, the attenuation properties of materials, the heating and cooling requirements of the machine and possible vibration isolation needs may all be required to arrive at a solution to a particular problem. Despite the need for experts, we can help in preventing people adopting stupid control methods. A sound insulator is needed to prevent noise travelling from one area to another. Dense, high-mass materials such as brick, concrete, heavy-gauge steel and hardwoods are good insulators. If we want to absorb noise and reduce its reflections from walls and so on we would use light, porous materials such as acoustic tiles and mineral wool. These are sound absorbers but are virtually useless as insulators and as expected, good insulators are poor absorbers.

Noise levels inside machinery enclosures will rise due to noise reflections from inside surfaces of the insulator. Noise reflected in this way is known as reverberant noise and is illustrated clearly in cathedral-type buildings. This reverberant noise can be reduced by lining the enclosure with an absorber such as mineral wool.

To be effective, enclosures must be air-tight – the smallest gap markedly reduces the attenuation given. In the design of enclosures we need to measure the sound level produced by the machine and the level of reduction required. When this is known, suitable insulators can then be chosen. The painting of absorbers will reduce their ability to absorb noise by producing an impervious noise reflective surface.

Noise and vibration are intimately linked and any vibrating surface will act as a noise source. Vibration isolation treatments such as the use of anti-vibration pads for machinery and flexible canvas collars between the fan and the ductwork in ventilation systems will help to reduce noise transmission along the ductwork.

Personal Protection

The following must be known when choosing hearing protection:

(1) the noise attenuation required to allow for acceptable noise levels at the ear;
(2) the frequency components of the noise source;
(3) the attenuation performance of the hearing protectors at various frequencies.

In the Code of Practice, the typical performance of fluid seal muffs is given along with mean and standard deviations which results in the calculation of the assured protection given by the muff for the majority of users. This example is reproduced in Table 5.3.

Table 5.3: Protection Given at Different Frequencies

Frequency (Hz)	125	250	500	1 k	2 k	4 k	8 k
Mean attenuation (dB)	13	20	33	35	38	47	41
Standard deviation (dB)	6	6	6	6	7	8	8
Assumed protection (dB)	7	14	27	29	31	39	33

From the table, it can be seen that the muff will give at least 7 dB protection at 125 Hz and 14 dB at 250 Hz, etc. Noise at higher frequencies is easier to attenuate than that at low frequencies (cf. 33 dB at 8 kHz and 7 dB at 125 Hz).

Different forms of hearing protection are available including ear muffs and re-usable and disposable ear plugs. In choosing protection, it is essential that the items chosen given adequate protection for the area being considered. This should be determined using the frequency analysis measurements and the appropriate information from the suppliers, so that the assumed protection levels can be calculated and judged as adequate. Wearers need to know why protection is required and how to use it properly. In issuing protection an element of choice should be given to the wearer, who will generally prefer one type mainly from a comfort point of view. It is important that the equipment issued should be robust in use and above all it should be worn. Hearing protection which stays in the locker will provide no protection whatsoever.

Audiometry

In terms of hearing loss, audiometry can be valuable in the detection of the first signs of occupational deafness but will not in itself prevent a loss occurring. Results obtained in such a test can give warning that measures must be taken to prevent further loss being sustained.

In audiometry, discrete frequencies at different sound levels are fed to ear muffs worn by the subject either in a quiet room or more usually in an acoustic booth. Both ears are tested individually and when the subject just fails to hear a signal at a specific frequency at a particular level, this is plotted on an audiogram similar to the one shown in Figure 5.5.

Figure 5.5: Example of Audiogram

2. Date....................Time since last exposure.........hrs.

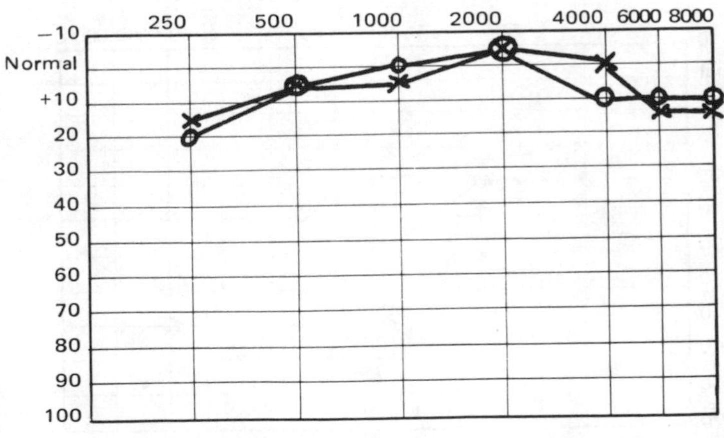

Operator:..O = R, X = L

When the test has been completed, one can see whether a deafness is occurring or whether an existing condition has worsened. Audiograms showing noise-induced deafness are shown in Figures 5.6 and 5.7.

Figure 5.6: Audiogram Showing Noise-induced Deafness

3. Date....................Time since last exposure.........hrs.

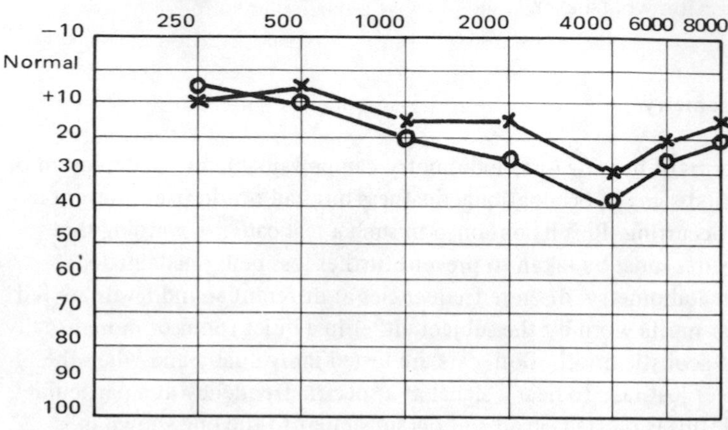

Operator:..O = R, X = L

Figure 5.7: Audiogram Showing Severe Noise-induced Deafness

4. Date...................Time since last exposure.........hrs.

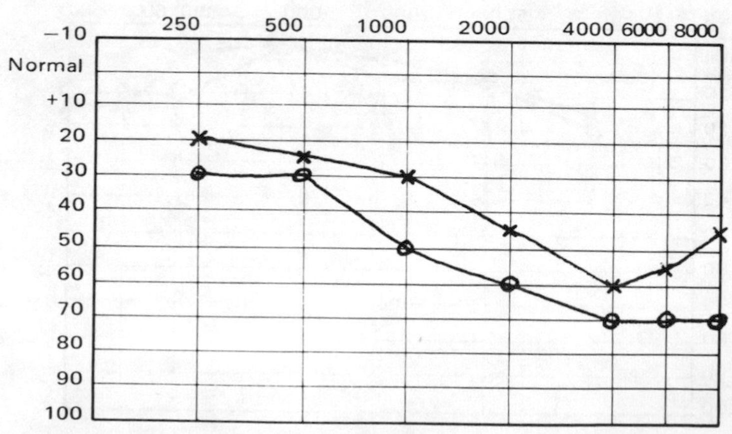

Operator:..O = R, X = L

Note the characteristic '4,000 Hz dip' associated with noise-induced deafness.

Audiometers can be of the manual or self-recording type and the choice of instrument depends largely upon the extent of the testing programme and staff availability. With the manual type, an operator controls the frequencies tested and the amplitude of the signal. The person under test merely signals that a particular sound is heard. Automatic testing relies upon the subject in that he alters the signal both in frequency and intensity when he fails to hear a particular signal. This is generally better where large numbers are being tested.

The following must be considered when carrying out an audiometric programme:

Instrument Calibration

A daily check should include an examination of the leads, earphones and switch positions. Stepwise increments in the signal intensity should be clearly heard and by checking the hearing of a known subject, for example one of the nursing staff, the reproducibility of the equipment can be assessed. In this type of check, the resulting audiogram should not deviate from earlier ones by more than 5 dB.

A prime calibration should be carried out at least annually and preferably twice a year by a specialist laboratory which will check the instrument against established British Standards.

Persons Tested

A temporary hearing loss will result from disease, a head cold or wax-filled ears. Audiograms should therefore only be carried out in the absence of these conditions, and preferably after quiet periods, for example after a quiet weekend, annual holidays or at least first thing in the morning. If not, then a temporary deafness may mask or exaggerate any permanent loss.

When

A pre-employment audiogram will indicate whether a hearing loss has been sustained in previous employment or in leisure pursuits such as shooting, attendance at discos, motor cycle scrambling, power boat racing and so on. To check the efficacy of a hearing conversation programme, audiograms should be repeated periodically, such as annually or six-monthly if work in a particularly noisy environment is undertaken.

Where

Tests should be carried out in a properly designed acoustic booth or at least in a quiet part of the factory (with background noise level of less than 40 dB at 500 Hz) preferably within the medical department. The room or booth should not be too far from the factory so as to allow the test to be carried out in a reasonable time.

In summary a hearing conservation programme should contain the following elements:

(1) Noise survey – to determine risk.
(2) Area designation – warning notices at the boundaries.
(3) Engineering control measures – designed by experts.
(4) Personal protection – chosen for its efficiency and acceptability.
(5) Worker education – films, booklets, instruction in the use of personal protection.
(6) Management commitment – 'enforcement' and example.
(7) Audiometry – to check on control procedures and minimise on 'deafness'.
(8) Periodic surveys and equipment changes.
(9) Repeated checks on (1)-(8).

Normally all OH team members are involved in this programme, e.g. the nurse will do the audiometry and education. However, if the nurse is not supported by a full OH team, then a consultant acoustic specialist will be required for the other aspects of the programme.

Communication Difficulties

Speech communication requirements can be assessed using noise rating numbers or increasingly today using dBA levels. Figure 5.8 shows the noise rating curves and to determine the noise rating number for a particular area the noise levels in decibels at the frequencies 500, 1,000 and 2,000 Hz are measured and plotted on the diagram. The noise rating number is then given by the highest curve in relation to our measurements. This is illustrated by the examples given.

Examples

		Frequency (Hz)			NR number
		500	1 k	2 k	
Noise level (dB),	Example (1)	60	70	70	75
	Example (2)	50	57	50	60
	Example (3)	48	40	40	45

Figure 5.8: Noise Rating Curves

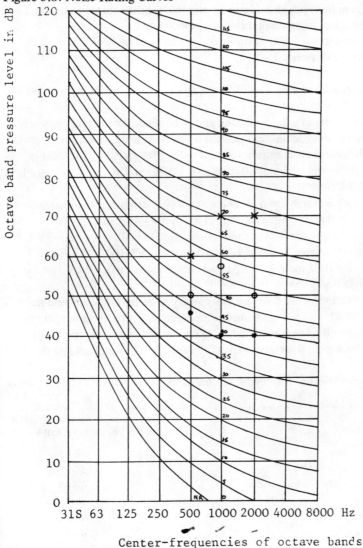

X = Example 1 ; O = Example 2 : o = Example 3

Source: Kosten, C.W. and van Os, G.J. 'Community Reaction Criteria for External Noises', Proceedings of a Conference on the Control of Noise (HMSO, London, (1962)

dBA levels are being used increasingly to define acceptable background noise levels and as a guide the relationship between dBA and NR numbers is given by applying the equation:

dBA = NR number + 5

In the examples given the dBA levels would therefore be 80, 65 and 50 dBA.

Difficulties in speech and telephone communication will obviously increase as the background noise level rises. Telephone conversation may be described as satisfactory up to 55 dBA (NR 50), increasingly difficult between 65 and 80 dBA (NR 60-75) and virtually impossible over 80 dBA (NR 75).

Speech will only be intelligible at specific distances under specific background levels. For example, at around 45 dBA intelligible conversation is possible at some 7 metres (21 ft) distance between the subjects. However, if the level rises to around 85 dBA our subjects need to be only 0.07 metres (2½") apart. If we shout one to the other the intelligibility distances double.

For various work areas we can specify maximum background noise levels which will not cause undue annoyance. These levels are generally well below the levels regarded as damaging to hearing. Examples of these background levels are given in Table 5.4. These should be viewed

Table 5.4: Recommended Maximum Background Noise Levels for Work Areas

Work Area	Maximum Background Noise Level (dBA)
Private office	40-45
General offices/plant offices	50-55
Workshop offices/plant offices	60-65
Workshop areas where communication necessary	70-75
First-aid area	40-50
Canteen	65-70

as rough guidelines only and complications arise if discrete frequency tones are present and if the background noise level is likely to mask warning signals.

In terms of annoyance, it is clear that noise emitted from a factory or any work location may result in complaints from the community. Due to the difficulty in applying a measurable scale to annoyance, attempts have been made in British Standard 4142 'Method of Rating Industrial Noise Affecting Mixed Residential and Industrial Areas' to evaluate the validity of complaints. In the standard, background noise levels for areas, such as rural and suburban and so on, are measured, or assigned if measurements are not possible as in the case of a factory which cannot be 'switched-off'. To this base figure various corrections are either added or subtracted to obtain a 'should-be' background level. By comparing this recommended value and the actual measured value that exists, it is possible to say whether the complaints received are valid. Generally, this happens when our recommended background level is greater than 10 dB below the level that exists while the factory is in operation. In these estimates, corrections are applied for the type of noise, that is whether a discrete frequency is noticeable and for the duration of the noise or its intermittency.

The Noise Abatement Act of 1960 takes the view that it is an offence to create a public nuisance and has been succeeded by the Control of Pollution Act of 1974 which is concerned with the control of noise emission. In both these, the local authority is empowered to act where noise amounting to a public nuisance exists. Part III of the Control of Pollution Act is concerned with all forms of nuisance and includes provision for the serving of notices on occupiers of premises requiring that noise reduction be instituted as well as limiting certain activities, for example, the use of loudspeakers in streets during certain hours. Additional regulations cover work on construction sites and noise from plant and machinery. The local authority can also designate noise abatement zones and noise measurements made by them in pursuit of this type of activity are kept in a noise register which is open to public inspection.

Noise Lesiglation

Noise Abatement Act 1960.
Control of Pollution Act, Part III 1974.
The Woodworking Machines Regulations 1974. Part x deals specifically with noise.

Summary of Key Points

(1) Excessive exposure to noise can produce a noise-induced

deafness in addition to the hearing loss due to ageing. Once deafened by noise there is nothing than can be done to restore hearing acuity.

(2) Noise can also produce communication difficulties and annoyance.

(3) Measurements of noise exposures (or noise dose) are relatively simple to carry out using a calibrated sound-level meter or personal dosimeter.

(4) The Department of Employment in its 'Code of Practice for Reducing the Exposure of Employed Persons to Noise' gives acceptable exposure levels – although the aim must be to minimise exposures wherever possible.

(5) Exposures can be reduced by:

   (a) reducing the noise level – quiet equipment, acoustic enclosures;
   (b) reducing exposures by the use of peacehavens and hearing protection (also by limiting entry to high-risk areas).

(6) Hearing conservation programmes should include an assessment of the noise levels, engineering control procedures, worker education, audiometry and the use of personal protection.

(7) Acceptable background noise levels for specific areas are known but annoyance is more difficult to define.

(8) A standard for defining community annoyance is available.

Further Reading

Department of Employment, *Code of Practice for Reducing the Exposure of Employed Persons to Noise* (HMSO, London, 1972)

Sutton, P. *The Protection Handbook of Industrial Noise Control* (Alan Osborne & Associates (Books) Ltd, London, 1974)

Webb, J.D. (ed.) *Noise Control in Industry* (Sound Research Laboratories, Sudbury, 1976)

6 THE THERMAL ENVIRONMENT

Introduction

Man needs to maintain an internal temperature of 36-38°C in order that the main organs can function normally. However, the body itself creates heat by metabolism and this *metabolic heat* increases with the rate at which work is being done. Because of this and because the ambient thermal conditions can vary enormously, the body has a regulatory system to control the internal temperature. When the metabolic heat is created at a rate that the body cannot remove into the environment the internal temperature increases and the control system becomes overloaded resulting in heat strain. Heat strain, of course, is a general term covering a multitude of physiological problems depending on the degree of exposure, but it can range from a simple loss of concentration (and performance) to death.

The purpose of this chapter is to provide sufficient background to allow the reader to recognise a situation where the thermal environment is potentially hazardous to health, to provide information on the important factors of that environment and how to measure them. The indices used as guidelines to gauge the extent of the hazard are also given, together with details of how to use the most commonly applied index.

Probably the most common failing of those involved in industrial health is to underestimate the importance of *measuring* the thermal environment. All too often it is felt that a quick 'walk-thru' survey will suffice. Inevitably the opinion is that a work area is 'nice and warm'. However, what is not recognised is that the subjective opinion – even of experts – of a thermal environment is based on the most recent past experience. For example, a nurse working in her medical unit might be called to a plant area suspected of being 'too hot to work in'. In order to reach this area she might have to leave her comfortable office and go outside to the main plant building. In winter she will experience a cold, chill wind. On arriving at the work area her most recent experience is 'cold'. She immediately gains a pleasure-response by walking into the warm area and her subjective opinion is biased. Only by staying in this area for some time and by producing the metabolic rate of the operators can the nurse reach the correct subjective opinion.

An important message from this chapter therefore is: *do not rely on your subjective opinion of a thermal environment – measure it properly and assess the rate at which the operators are working.*

Health Effects

The human body shows acute effects as a result of heat, e.g. sweat, as well as longer-term effects such as sunburn. Breakdown of the heat balance occurs when the control centre, the *hypothalamus,* at the base of the brain, cannot maintain the required heat balance by sweating, etc. When this control system fails, a form of heat illness occurs. The form of heat illness can vary enormously with the degree of stress applied but the following conditions are typical.

(1) Heat oedema – 'deck ankles' resulting from venustasis (the ankles swell).
(2) Heat syncope – dilation of the blood vessels.
(3) Cramps – loss of water/salt balance.
(4) Hidromeiosis – sweat glands cease to function.
(5) Prickly heat – blocked sweat glands resulting in sweat escaping into the dermis.
(6) Anhydrosis – excess radiant heat.
(7) Heat hyper-pyrexia – body temperature elevates to 39-40°C (prior stage to heat stroke).
(8) Heat stroke – body temperature up to 42°C with anhydrosis and signs of neurological failure (ultimately coma).

The Important Factors to Measure

The factors that govern whether a thermal environment is hazardous to health are those which control the heat exchange between the body and the environment. These are: conduction, convection, radiation and evaporation. It should become obvious at this point that the assessment of a thermal environment by means of a single mercury thermometer is impossible – it only represents one part of the jigsaw. Let us examine these factors in a little more detail to see what effects may influence them.

Conduction. For the body to lose or gain heat by conduction it must be in contact with a solid surface, e.g. an operator leaning on a hot or cold surface. This is normally not the case and therefore conduction is usually not important in our situation.

Convection. This relies on the heat being transferred on a current of moving air. The amount of heat removed depends on both the amount of air movement and the difference between the temperatures of the air and the skin. A hot body relies a lot on losing heat in this way, but if it is fully clothed the amount of air movement next to the skin is very small indeed. Immediately, therefore, we can define three important factors to be measured, i.e. the wind speed, the amount of clothing and the air temperature.

Radiation. If the environment contains sources of radiant heat, e.g. electric fires or furnaces, then what is called the mean radiant temperature of the surroundings will be high. If it is higher than the surface temperature of the body (i.e. the skin) then heat will be transferred to the body by radiation. We now, therefore, have another important factor to measure, i.e. the mean radiant temperature.

Evaporation. When the body sweats it does so to lose heat. The heat is used up in evaporating the sweat but the sweat will evaporate at a rate that depends on how much moisture is already in the air – i.e. the humidity of the air. Evaporation is also speeded up if *fresh* air is passed over the body, so that the air immediately around the body does not become too moist and cannot absorb any more water vapour. The effect of air speed again comes into play. For evaporation, therefore, the important factors to measure are humidity and wind speed. These factors refer to the environment but let us not forget the operator who will be producing heat himself, by working (metabolic heat).

Before moving on, we can now list the important factors that need to be measured to define the nature of the thermal environment:

(1) wind speed (air movement);
(2) clothing index (how much clothing is worn);
(3) mean radiant temperature;
(4) humidity;
(5) work-rate;
(6) air temperature.

Measurement of the Thermal Environment

First, it is important that the selection and placement of the instruments are done correctly. For this reason the reader should ensure that the mechanisms discussed briefly above are fully understood. Only by the investigator realising why a particular instrument is being used can he or she hope to use it correctly. If any doubt exists at this stage then reference should be made to a fundamental physics text (see further reading).

Secondly, the investigator must become familiar with the instrument itself. Each device has its own particular mode of operation. This is even more true of the thermal measuring devices since they need time to equilibrate and readings should only be taken when the investigator is sure the reading is stable.

(1) Air Temperature Measurements

Air temperature can be measured by a variety of instruments. For reference, three main types are listed here.

(a) Mercury (or Alcohol)-in-Glass Thermometer. This is the most common but potentially the most inaccurate. Care must be taken to calibrate over the required range using a temperature-controlled liquid (oil or water) and a known standard such as a certified thermometer. In use, several minutes must be allowed for the thermometer to reach equilibrium before reading.

(b) Thermocouple Thermometer. This requires a potentiometer for read-out and again requires calibration. It takes much less time to reach equilibrium than the mercury-in-glass type and can be read remotely – depending only on the length of wire used. On the negative side, it can be bulky and cumbersome and tends not to be used for air but for surface (skin) measurements.

(c) Thermistor Thermometer. This is much simpler than the thermocouple device because it does not require balancing to obtain a reading. It relies solely on the change of electrical resistance with temperature. Read-out equipment is powered by battery and is portable – making it ideal for field use. Again, like the thermocouple, it can be read remotely. This instrument is gaining in use due to its robustness, portability and quick response.

No matter which of these instruments is used, the investigator should take care that it is the *air* temperature that is being measured. If radiant exchange is occurring due to walls or surfaces being warmer or cooler than the air, then the thermometer sensor should be shielded. This is done simply with heavy aluminium foil wrapped around the sensor, e.g. bulb in the case of the mercury-in-glass thermometer; however, the foil should be wrapped loosely so that it does not prevent the flow of air around the sensor. As a final point, in air temperature measurement, it should be remembered that most work areas are not square. If a room is asymmetric, e.g. L-shaped, the investigator should take care to read air temperatures at several points to estimate the average air temperature – always remembering that the most critical area is where the man is working.

(2) Air (Wind) Velocity Measurements – Anemometers

Air movement is seldom in one direction only. The investigator should take care, therefore, that the device used to measure air movement is not uni-directional, i.e. measures movement in one direction only. These uni-directional meters are in common use in the assessment of ventilation systems but are of no use in thermal environment studies.

Two main types exist which are commonly used in field studies of the thermal environment, namely:

(a) Thermoanemometers. These rely on the effect of the wind on the rate of cooling of either a heated resistance wire or heated thermocouple (see Figure 6.1). As the velocity of the air increases, so the respective electrical circuit balance changes and the velocity is read directly on a meter. They are portable, battery-powered and give an instantaneous response. Because the wire or bead requires supports to hold it in place, these can restrict air movement across the sensor and the probe should be rotated to give the maximum reading. The speed of response and portability of these devices has resulted in their becoming extremely popular of late.

(b) Kata Thermometer. Basically this is an alcohol-filled thermometer with an extra-large bulb. The bulb is immersed in warm water until the alcohol column rises into the upper reservoir and it is then removed from the water and wiped dry. Suspending the thermometer in the air causes the movement of air to cool the alcohol – the column of which slowly falls. By timing the fall between the two marked points on the stem, the air velocity can be obtained from nomograms provided with the instrument. This thermometer is obviously made in several ranges

Figure 6.1: Standard Hot-wire-bead Anemometer. Note the rubber cover for the head whilst the instrument is not in use and also the facility for an extension arm.

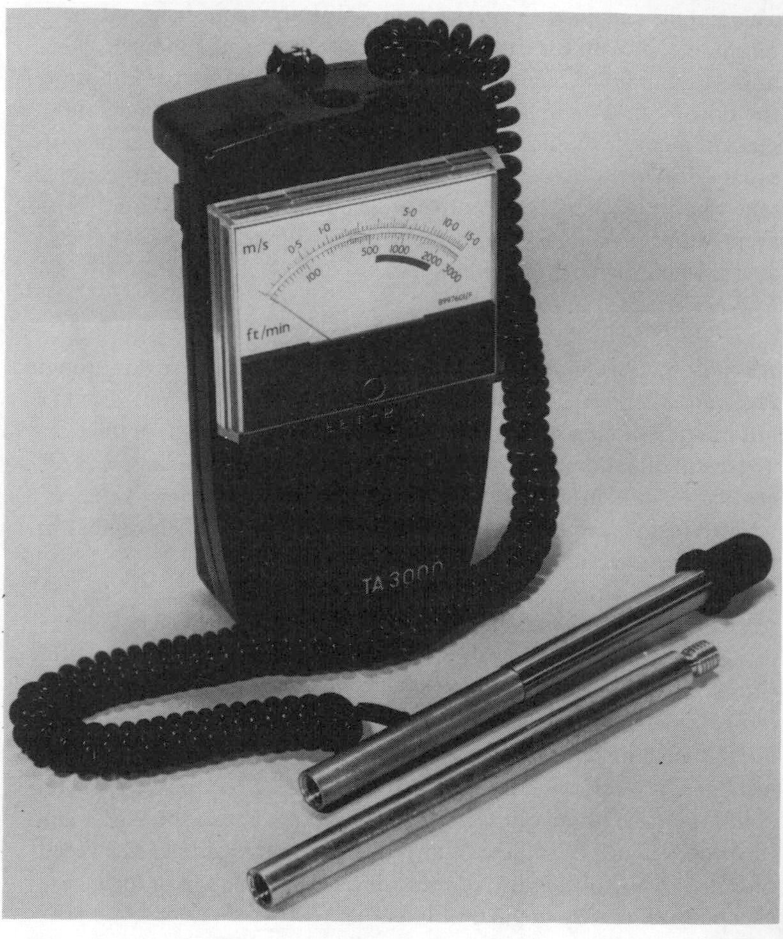

but each will have the bulb silvered to prevent radiant heat exchange. It is time-consuming and cumbersome (always needing a thermos of warm water handy) but is accurate at low wind-speeds – say less than 10 metres per minute in which range the thermoanemometer can be less sensitive.

As with all measurements in this field, care must be taken that the reading represents the area as a whole. Investigators should use their judgement to ensure that if concentrated draughts exist they do not read these as existing throughout the whole room. The readings should also be taken at arm's length so that interference from the body is minimised.

(3) Humidity Measurements – Psychrometers

Most of the indices used to assess the thermal stress of an environment utilise the wet and dry bulb temperature readings as indications of humidity. We shall not, therefore, concern ourselves with direct reading instruments of relative humidity – such as the hair hygrometer – but restrict this section to a device that measures the temperature variables, i.e. the *sling hygrometer* (see Figure 6.2).

If a normal mercury-in-glass thermometer has its bulb surrounded by a cotton wick soaked in water, the water will evaporate into the air to an extent depending on how much water is already present in the air, i.e. the humidity. This evaporation needs heat and this heat is taken from the thermometer bulb, with the result that the bulb cools. This degree of cooling gives a measure of the humidity.

To ensure that the air above the cotton wool does not become saturated with moisture, the whole thermometer is rotated. Since a reading of the air temperature itself is also needed, a *dry bulb* thermometer is incorporated alongside. This is achieved by putting both thermometers in a *sling* similar to a football supporter's rattle. A water reservoir is also incorporated to maintain the wick of the *wet bulb* always moist.

After rotating this sling for about one minute at arm's length, both temperatures are quickly taken. The wet bulb should be read first since once the sling has stopped the wet-bulb temperature will begin to rise again. This procedure should be repeated until constant readings are obtained.

For the purposes of thermal stress indices the humidity itself is not required, since defining the wet- and dry-bulb readings will define the humidity and these alone are used in the indices. For information, however, the humidity and relative humidity can be read from a psychrometric chart from the two temperatures. Unless the air has 100 per cent relative humidity, i.e. is completely saturated with water vapour, the wet-bulb temperature will be less than the dry-bulb temperature.

Practice with this instrument is essential so that the investigator can gauge for how long the sling requires rotating before a stable reading is taken. In any event, at least three readings should be taken until a constant reading is achieved.

Figure 6.2: Rotating-sling Hygrometer for Wet- and Dry-bulb Thermometer Readings

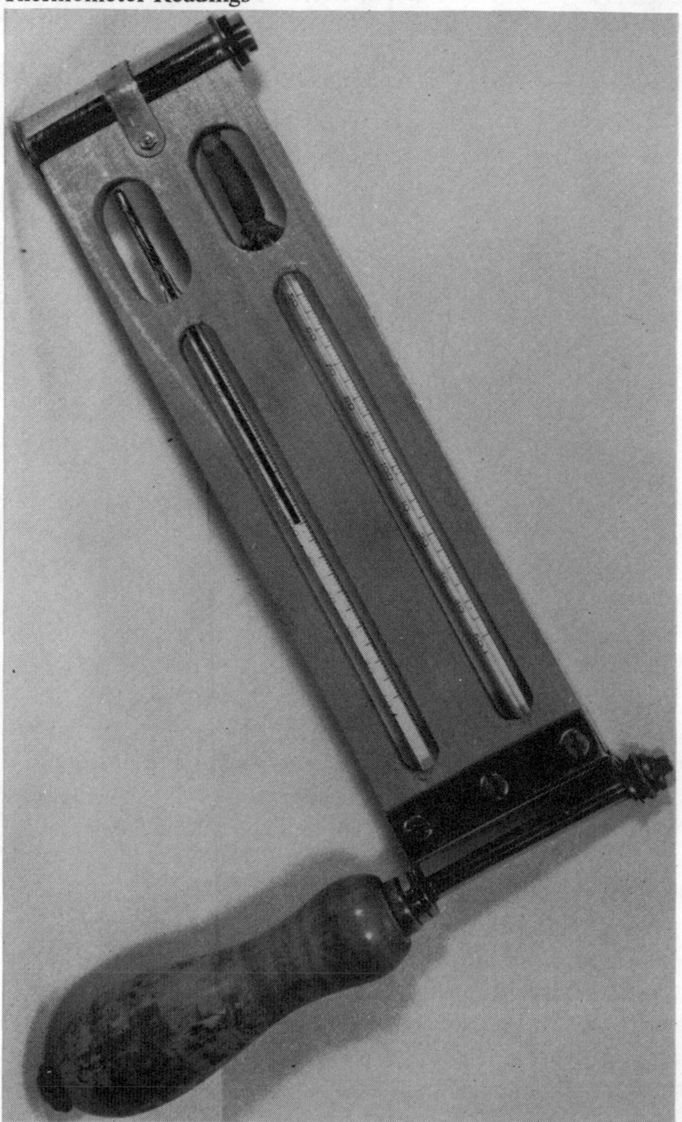

(4) Radiant Temperature Measurement

The mean radiant temperature is measured by using a blackened globe called a *Vernon Globe* (see Figure 6.3). This is a copper sphere 15 cm in diameter, of which the outside is painted matt black. A thermometer is

Figure 6.3: Vernon Globe Thermometer for Radiant Temperature Measurement

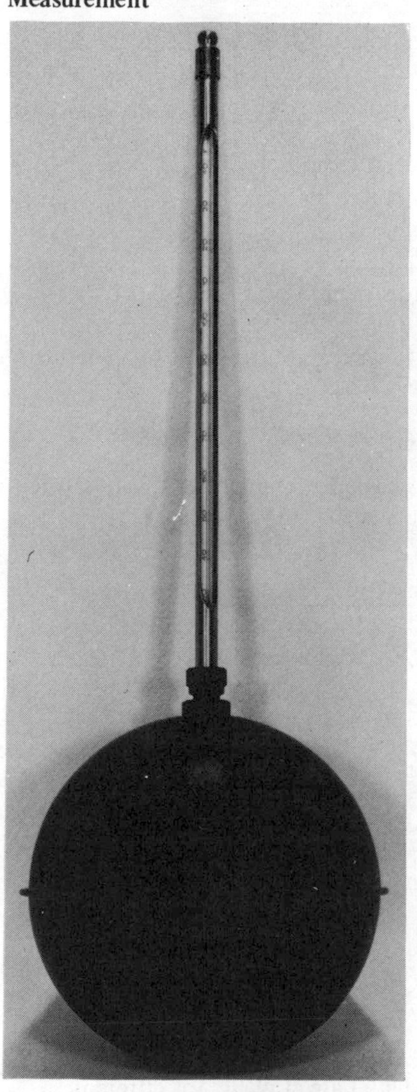

inserted so that the bulb is located in the centre of the globe.

The globe, with the thermometer firmly attached, is hung freely at the point of measurement. Normally, at least 20 minutes is required for thermal equilibrium before the thermometer reading is taken.

Some indices use the globe temperature itself, but others require the mean radiant temperature of the surroundings. This can be calculated from the equation:

$$Tw^* = 100 \left(\frac{Tg^*}{100}\right)^4 + 2.48V\,(Tg - Ta)$$

where Tw* = mean radiant temperature in Kelvin (273+ °C)
Tg* = globe temperature in Kelvin
Tg = globe temperature in °C
Ta = air temperature in °C
V = air velocity in metres/minute.

The most commonly used indices, however, use the globe temperature only.

(5) Clothing Index and Work Rates

These are well tabulated and one needs only reference to the literature. As a guide, Tables 6.1 and 6.2 should suffice most needs.

Table 6.1: Work Rates

Activity	Metabolic Rate (kcal/hour)
Sitting	80
Standing	100
Walking (4 km/hr)	220
Standing : light hand-work	160
Standing : heavy hand-work	200
Standing : light arm-work	270
Standing : heavy arm-work (e.g. sawing)	360-580
Work with whole body:	
light	270
moderate	360
heavy	480

The heat stress indices discussed later are available for different clothing indices. In practice, the investigator will utilise the 'normal' index which relates to light summer clothing, i.e. the usual for workers involved in any degree of activity. However, the other clothing indices are shown in Table 6.2 merely as a reference to indicate that the clothing factor has been considered by the developers of the indices.

Table 6.2: Clothing Indices

Clothing	Clothing Index
Nude	0
Shorts only	0.1
Shorts, open-necked short-sleeved shirt, socks and sandals	0.35
Light summer clothing	0.5
Business suit	1.0
Polar weather suit	3-4

Indices of Thermal Stress

Being involved in industrial hygiene, the reader will understand the desire to assess a potential health hazard in terms of a single number, which can then be related to guideline limits such as the Threshold Limit Values. With thermal stress the need is the same but the difficulties are magnified by the complexity and number of the variables. Of course, thermal stress has been known of for many years – particularly in the armed services where forced marches through alien climates are the rule rather than the exception. Not surprisingly, therefore, a great deal of the foundations of the now commonly used indices came from medical teams of the armed forces. As a result they are empirical in nature, based on the response of subjects exposed to controlled environments.

Even working on an empirical basis, there are different approaches in actually measuring the response of a subject to a particular environment. First, the researcher can make physiological observations of the subject, i.e. heart rate, respiration, sweat rate, etc. Secondly, the researcher could do an analysis of the physics involved, i.e. measure the amount of heat loss by sweating, etc. Thirdly, the researcher could merely ask the subject for a personal opinion, i.e. a subjective response.

All three methods have been tried and the intention here is to give one example of the indices produced from each of these approaches. Many others exist but the following three represent the most commonly used indices for assessment of thermal stress.

(1) Index Based on Physiological Observations: Wet-Bulb Globe Temperature Index (WBGT)

Developed originally for armies on active service in the desert, the

WBGT has since become established in industry and is now incorporated into the American Conference of Governmental Industrial Hygienists (ACGIH) *Threshold Limit Values for Physical Agents in the Workroom Environment* – published annually – as recommended practice. As defined in the ACGIH book, WBGT values are calculated by the following equations:

(a) Outdoor Work with a Solar Load

$$WBGT = 0.7WB + 0.2GT + 0.1DB$$

(b) Indoor Work or Outdoors with *no* Solar Load

$$WBGT = 0.7WB + 0.3GT$$

where, WB = Natural wet-bulb temperature
(this means that the wet-bulb thermometer reading on the sling hygrometer is taken *without* rotating the sling, i.e. it is not ventilated forcibly but *naturally*)
DB = Dry-bulb temperature
(as described earlier, this is the air temperature)
GT = Globe thermometer temperature.

Having calculated the WBGT, the number in °C is then compared to guideline limits of work: rest regimes from either a table or graph (Table 6.3 and Figure 6.4).

Table 6.3: Maximum Permissible WBGT Readings

	Work Load		
Work: Rest Regime	Light	Moderate	Heavy
Continuous work	30.0	26.7	25.0
75% Work: 25% Rest – each hour	30.6	28.0	25.9
50% Work: 50% Rest – each hour	31.4	29.4	27.9
25% Work: 75% Rest – each hour	32.2	31.1	30.0

An example of this is that if the WBGT in the work area is 30°C then the worker has three options:

(1) Can do continuous light work.
(2) Can do slightly less than moderate work for 50 per cent of each hour and rest the remainder.
(3) Can do heavy work for 25 per cent of each hour and rest the remainder.

Figure 6.4: WBGT Values Plotted Against Rate of Work for Various Work: Rest Regimes

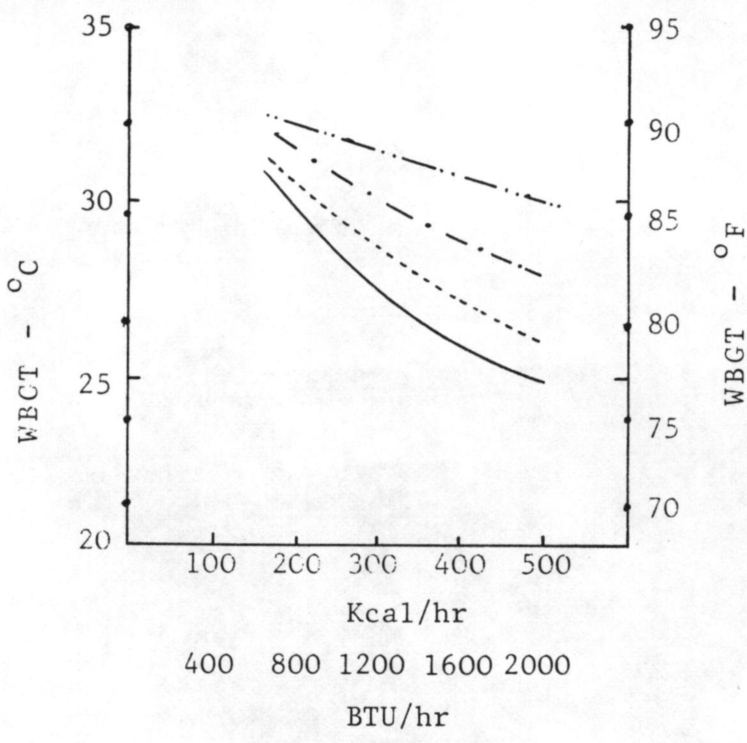

Source: Threshold Limit Values for Chemical Substances & Physical Agents in the Workroom Environment with Intended Changes for 1980 (ACGIH).
Key:

———————— continuous
— — — — — — — 75% work – 25% rest each hour
———— · ———— 50% work – 50% rest each hour
———— · · ———— 25% work – 75% rest each hour

The WBGT is certainly recognised as the simplest and most suitable index at present since it incorporates all the relevant variables, which can be measured on a single instrument (see Figure 6.5). Note that although wind velocity is not physically measured, it is incorporated via the natural wet-bulb temperature since this reading will be dependent on the air movement around it.

Figure 6.5: WBGT Instrument Giving Facility for all Necessary Readings for this Index

(2) Index Based on Analysis of Heat Exchange: Heat Stress Index of Belding and Hatch (HSI)

Although revisions have been made to incorporate a more realistically clothed worker, the original concept of this index remains the same as devised by Belding and Hatch. Essentially two quantities are estimated from the environmental and metabolic rate data. These are: (1) the *required* evaporative heat loss (by sweating) to achieve heat balance (Ereq) and (2) the *maximum* evaporative heat loss possible in that environment (Emax).

The ratio of Ereq:Emax is then calculated and related to an

allowable exposure time. Obviously, if Ereq is less than or equal to Emax then the worker can work continuously without any ill effects. When Ereq is greater than Emax then heat will build up in the body and only a limited time for working is allowed – based on a maximum allowable increase in heat load in the body.

The measurements required for this index are:

(1) globe temperature;
(2) air temperature;
(3) wet-bulb temperature;
(4) air velocity;
(5) metabolic work-rate.

It is beyond the scope of this text to go through the nomograms involved in the calculation of Ereq and Emax but these can be found on page 421 of the NIOSH book *The Industrial Environment – Its Evaluation and Control.* An example calculation is also given for those wishing to pursue the use of this index.

(3) Index Based on Subjective Preference: Effective Temperature (ET) and Corrected Effective Temperature (CET)

In the original form the ET made no allowance for the radiant temperature which was assumed to equal the air temperature. The inclusion of the radiant heat factor was applied later and formed the new CET. Two different nomograms are available for two different clothed states: semi-nude and light summer clothed (normal work clothes). Measurements necessary are:

(1) wet-bulb temperature;
(2) wind velocity;
(3) globe temperature or air temperature, if same.

The CET found from the nomogram (in °C) is then regarded as the temperature of a still, saturated environment that provides the same subjective response in individuals exposed to the environment under examination.

The World Health Organisation recommends that the following limits should apply: sedentary work, 30°C CET; light work, 28°C CET; heavy work, 26.5°C CET. Because several nomograms exist they will not be included here, but page 419 of the NIOSH book mentioned in the previous section gives an example for those wishing to pursue this further.

As a conclusion to this section it is worth noting that the data generated from these three indices should provide similar *guidelines* to limit worker exposure to thermal stress – although data themselves will not be identical. Since the WBGT is the simplest to use (involving no nomograms) it is the one recommended. In order to simplify this further, in the early 1970s James Botsford developed an integrating thermometer that essentially gives a direct reading of the WBGT. This is a wet-globe instrument based on a small copper sphere (6 cm diameter) fitted with a black cotton wick and water reservoir (see Figure 6.6).

Figure 6.6: Botsball Thermometer Giving an Approximation of the WBGT

While the globe integrates the air temperature, radiant temperature and air movement, wetting the sphere introduces the fourth variable, humidity. The thermometer reading, therefore, approximates the WBGT. Other devices exist but usually these merely provide electronic solutions to the equation after feeding in the relevant data directly from the thermocouple or thermistor thermometers. Faced with many alternatives, the reader must judge each on its own cost/benefit ratio.

Solving Thermal Stress Problems

Having defined a situation of thermal stress the investigator will undoubtedly be asked to solve it. The control measures that are selected will be based on the variable that most significantly contributes to the stress. To reiterate, the important variables are:

(1) rate of work;
(2) time and frequency of work;
(3) heat exchange with the environment by convection, radiation and sweat loss;
(4) clothing index.

The first lesson to be learned is that there is never one solution that always works. However, altering (1) or (2) will usually provide problems from the viewpoint of industrial relations and also increase product costs. Altering (3) will involve capital and product costs, whereas the possibility of changing (4) will be unlikely, since the subject will probably be down to the 'core' minimum by the time the investigator is called in. Few friends are to be won in this game. If it is assumed that altering (3) is the solution which is less fraught with obstacles, then life becomes a little easier.

The best solutions arise from a commonsense approach. For example, if the environment is hot and wet, i.e. of high humidity, then increasing the air movement by a large fan will obviously only produce a small improvement. However, if the air movement was increased *and* the humidity lowered then large improvements will occur because sweat will be able to evaporate much more easily. If the problem is in a hot and dry environment, then increasing the air movement alone will provide large increases in heat exchange. If the air movement is already high, then removal of clothes or reduction of air temperature become the other options.

Radiant heat (e.g. from furnace walls) is a particularly difficult problem, since no amount of air movement will help. The answer here is to *shield* the worker from the radiant heat source. This is normally achieved by a finished aluminium sheet and using infra-red reflecting glass for any visual needs required. It is worth noting here that the polished-shield principle only works if the shine is maintained. Dirty shields do not work.

The above brief outline of control measures is intended only to indicate that unlike many industrial hygiene problems the control of thermal stress can be relatively easy and needs only a commonsense approach.

Low Temperatures

Thermal stress implies both hot and cold stress. This chapter has concentrated on heat stress primarily because cold stress is rarely a problem in industry. In situations where it is necessary, insulation can be increased by extra clothing and usually the only real problems are the head and the extremities. A drop in the central temperature is rarely a problem because the extra clothing and metabolic heat rate maintain the temperature easily. It is not recommended that sedentary tasks are carried out under cold conditions. It can be shown in fact that at 8°C and 0.1 m s^{-1} wind (calm) a sedentary person needs polar clothing to maintain comfort.

The key point here is that air movement has a large effect on the rate of heat loss from the body and clothing should be wind-resistant for workers in 'cold-store facilities' for example. With well-insulated head and extremities and wind-resistant clothing there should not be a problem in this area.

Legislation

In the UK the 1963 Offices, Shops and Railway Premises Act made a provision that:

> a reasonable temperature be maintained in every room in which people are employed to work otherwise than for short periods. For rooms where a substantial proportion of the work does not involve severe physical effort a 'reasonable temperature' shall be not less than 60.8°F after the first hour. In all workrooms there must be an effective and suitable means of ventilation by the circulation of adequate supplies of either fresh or artificially purified air.

This, of course, is referring to 'comfort' conditions rather than thermal stress. There is in fact no specific legislation on heat stress. However, with the Health and Safety at Work Act, 1974 the fact that the employer must do everything 'reasonably practicable' to ensure the health of the workforce effectively enables the ACGIH guidelines on the WBGT index to be taken as a legal requirement – as long as it is 'reasonably practicable'.

Summary of Key Points

(1) Man needs to regulate his internal temperature within 36-38°C. The inability to do this results in thermal strain and potential death.

(2) Thermal stress exists when the heat produced by the metabolic work rate cannot be exchanged into the environment.

(3) The important factors to measure in a thermal environment survey are:

Wind speed
Air temperature (dry bulb)
Wet-bulb temperature
Radiant (globe) temperature
Work rate
Clothing index.

(4) Important measurement tools are:

Sling hygrometer
Vernon globe thermometer
Anemometer (wire or bead)
Kata thermometer
Shielded mercury-in-glass thermometer.

(5) Key heat stress indices are:

Wet-bulb globe temperature (WBGT)
Corrected effective temperature (CET)
Heat stress index (HSI)

(6) Control of heat stress *environments* can be achieved by altering the variable most affecting the heat exchange from body to environment i.e.:

Increase air movement
Lowering air temperature

Decrease humidity
Shield radiant heat sources.

(7) Cold stress can be avoided by use of wind-resistant clothing and insulating the extremeties.

(8) Specific legislation only exists for comfort at low temperatures (60.8° C) but 'HASAW' Act implies compliance with WBGT guidelines in ACGIH Threshold Limit Values.

Further Reading

General

Brief, R.S. *Basic Industrial Hygiene – A Training Manual* (AIHA, Ohio, 1975)

Davies, C.N., Davis, P.R. and Tyrer, F.H. *The Effects of Abnormal Physical Conditions at Work* (Churchill Livingstone, Edinburgh, 1967)

NIOSH *The Industrial Environment – Its Evaluation and Control* (US Government Printing Office, Washington, DC, 1973)

Specialist

Fanger, P.O. *Thermal Comfort* (Danish Technical Press, Copenhagen, 1970)

Kerslake, D. *The Stress of Hot Environments* (Cambridge University Press, Cambridge, 1967)

7 LIGHTING

Introduction

Artificial lighting allows us to continue with work at times when early cave-men would have given up hunting for the day. Even during daylight hours sufficient light for particular tasks fails to be provided by nature and, again, we are forced to rely on man-made light. To an increasing extent control rooms on chemical process units are built to be blast-proof and as a necessary part of their design they are 'windowless'. In these situations the dependence on artificial light is total and daylight hours fade into obscurity.

If we have to rely on light from sources other than the sun then it is important that we maximise the benefits obtained and minimise any problems that may occur. In this respect we can control our own man-made sun.

Health Effects

From observing a rainbow or even a drop or two of oil on the surface of a puddle, one can see that light consists of a number of individual colours which normally blend together to give normal, white light. This so-called visible range of the electromagnetic spectrum is bounded on one side by the infra-red spectrum and on the other by the ultra-violet spectrum. In terms of health effects visible light is relatively unimportant – although it is obvious that if insufficient light is present then we may fail to see clearly and trip over objects in our path. Again, if lighting levels are poor we may find ourselves feeling 'shut in', particularly in factories having a poorly-lit roof space, or we may find complaints of eye-strain occurring. These complaints are difficult to quantify and good general lighting certainly gives a working area a degree of openness that adds to the general well-being of the operators. On the other hand, both ultra-violet and infra-red radiation produce definite hazards to the eyes (see Chapter 8).

Visible Lighting

Light is essential for sight. Unless well-designed for the task in hand, artificial light can reduce visual performance. In the main the finer the task, for example minute electronics assembly, the greater the amount of light required for satisfactory performance. The amount of light emitted by a source or received by a surface is known as the luminous flux (measured in lumens). If we want to define the density of light (luminous flux) per unit area of surface, we use the term illuminance, the density of which is measured in lux (that is the illuminance per square metre of surface area) or the foot candle (the illuminance per square foot of surface area or the amount of light from one candle falling on one square foot of surface area). To convert from foot candles to lux, multiply by 10.76.

Glare

Glare may cause either discomfort (*discomfort glare*) or may reduce the ability to see the task properly (so-called *disability glare*) and occurs when some parts of the visual field are excessively bright in relation to the general level of brightness. The degree of glare is dependent upon the surface viewed, the illumination level, the position of the light sources within the visual field and the average brightness of the surroundings. Discomfort glare can be reduced or eliminated by reducing the contrast between the object viewed and the surroundings. Matt finishes on work tops and machines etc. are preferable and the position of the light source should be kept outside the normal field of vision. The contrast ratio between the task and the surroundings should be 10:3:1 as shown in figure 7.1. An example would be the use of a buff-coloured blotter between a black-topped desk and white writing paper.

It may be impossible to view parts of the task due to disability glare, for example when light reflects from shiny steel surfaces. Sometimes disability glare can be desirable, for example with security lighting directed towards boundary fences. People at the perimeter cannot see behind the lights whereas those behind will have a clear view of the fence. Disability glare can be reduced by having matt surface finishes or by altering the position of the viewed object, or the light source or both, in the visual field.

As stated before, surfaces (tasks) are seen due to their contrast with the surroundings and the light reflected from them. The latter can be increased by increasing the incident light falling on the surface and it is, therefore, clear that higher illumination levels allow objects to be seen with greater ease. The Illuminating Engineering Society in their 'Code

Figure 7.1: Contrast Ratios

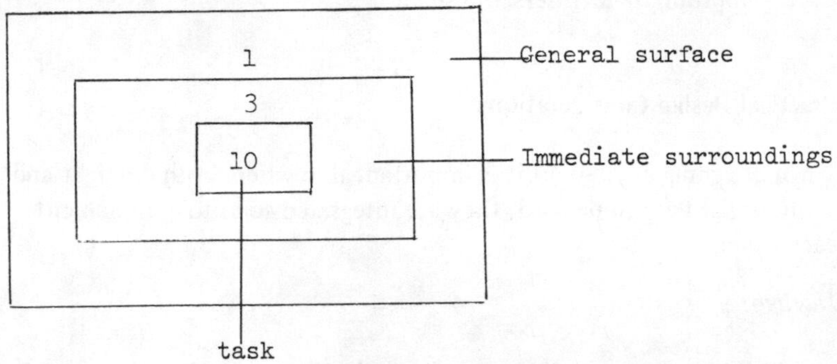

for Interior Lighting' recommend illumination levels in lux for a wide variety of tasks and also accommodation areas. Surface finishes in working areas should be chosen to give satisfactory reflectance levels. Table 7.1 summarises these recommendations.

Table 7.1: Recommended Colours and Finishes for Various Areas

Area	Reflectance (%)	Colour/Finish
Ceilings	70	Matt white or near-white
Walls	40-70	Light pastel colours
Floors	30	Light in colour, bearing in mind practical considerations; a high gloss finish is undesirable
Furniture	20-50	Light bench finishes

To overcome glare we can:

(1) Displace the source of glare out of the line of vision.
(2) Position lights correctly. Illuminate desks from the side with fluorescent tubes running parallel to the side edge of the desk and not the front edge.
(3) Desk tops and bench surfaces should be fairly light in colour. Black-topped desks should not be used.
(4) Avoid gloom in roof areas. This is particularly important with high ceilings. Increase brightness of the surrounding and furnishings.
(5) Choose the right reflectance for other surfaces (see Table 7.1).

(6) Fluorescent tubes and bulbs should be suitably screened with appropriate diffusers and shades.

Practical Design Considerations

When designing any layout it is important that where both daylight and artificial light are to be used, they are integrated so as to complement each other.

Daylight

(1) Windows and roof glazing should be chosen to give adequate light.
(2) Co-ordinate layout of machinery to maximise use of natural light.
(3) Distribute glazing evenly so as to allow an even distribution of light.
(4) Daylight from only one source should be avoided.
(5) Floors, ceilings and walls should have correct reflectance.

Artificial Light

(1) As (2) and (5) above.
(2) Distribute general background light evenly over working areas.
(3) Supplement general with localised lighting as required.
(4) Distribute lighting taking account of structure such as roof supports, etc.
(5) Allow overlap of one light pool with another so as to avoid dark spots.
(6) Spacing between fittings is related to the height above the task.
(7) Choose correct light types, bearing in mind:
    (a) level of illumination required;
    (b) colour rendering required from the lights (particularly important in textile and paint-spray work areas);
    (c) economic factors – cost of installation, lamp life, lamp efficiency (fluorescent tubes more efficient than filament bulbs), etc.
(8) Assess heat output from lights which may require extra ventilation.

Due to flicker from fluorescent tubes as a result of the alternating nature of the current, care must be taken to avoid the stroboscopic effect where moving machinery, e.g. lathes, may appear to be rotating more slowly or may even appear to have stopped. This can be eliminated by having the lamps out of phase, by screening the ends of the tubes or by using tubes with a long afterglow.

Colour rendering of light sources should be considered particularly in the textile and dyeing industries.

Measurement

In the main we are primarily concerned with the measurement of illuminance on the task and this is easily measured using a light meter having an appropriate photocell and scaled output which gives a direct reading in lux (see Figure 7.2). By comparing the levels found

Figure 7.2: Standard Luxmeter. The holed attachments allow range changes by limiting the area of the illuminated surface of the instrument.

with those recommended by the Illuminating Engineering Society in their Code of Practice, one can determine whether a particular situation is satisfactory. The light meter is placed on the task/bench and the illumination level noted. Care must be taken to ensure that no unusual shadows are cast but, at the same time, one should not overestimate the light falling on an area by removing shadows that are normally present.

For example, the illumination level at the keyboard of a typewriter may normally be overshadowed by the typist. If that is the case, measurements should be made with the typist in position.

Legislation

The Factories Act 1961, Section 5 requires that 'sufficient and suitable lighting' shall be provided. Similar requirements are made in the Offices, Shops and Railway Premises Act of 1963.

Summary of Key Points

(1) Need sufficient lighting to see the task in hand.
(2) Badly designed installations may result in complaints of glare.
(3) Low levels of lighting may result in complaints of eye strain and may also produce trip hazards.
(4) Colour schemes should be chosen to give adequate reflectance of light.
(5) Light sources should be chosen so as to avoid colour distortion in work areas where this is important.
(6) Lighting levels at the workplace are easily measured using a light meter and results compared with the standards set by the Illuminating Engineering Society.

Further Reading

Illuminating Engineering Society *IES Code for Interior Lighting* (The Illuminating Engineering Society, London, 1977)

Lyons, S. *Management Guide to Modern Industrial Lighting* (Applied Science Publishers, Barking, Essex, 1972)

8 ELECTROMAGNETIC RADIATIONS

Introduction

In the modern world the concepts of atomic energy and radiation are known to all and feared by most. The layman understands now that tremendous energies are involved in holding together even the simplest atom of hydrogen and that when these atoms are 'split', the energies can be released in the most devastating manner. The reason for this is that the energy contained in the atom, used to hold it together, can be transformed into heat, light and other forms of radiation – just the same as electrical energy fed to an electric fire will be transformed into heat, light and infra-red radiation. The point to be made here is that the effects that are created e.g. visible light, infra-red radiation, ultra-violet, micro waves, etc. are all forms of the same 'energy' – just released in a different form.

The 'energies' are in fact released in a particular manner that controls what form they take. Throwing a stone into a lake will create a wave that moves outwards from the point at which the stone hits the water. The energy created by the stone hitting the water is released in the form of waves. The size and frequency of the waves (how many waves pass a point in a given time) depends on how large the stone is and the force with which it hits the water. In just the same way, the energy released from an atom is released in the form of waves. The length and frequency of these waves depends on how much energy the atom is releasing. It is the length and frequency of the waves that controls what form of energy we call it and the effect it can have on the human body.

The various energy types therefore, are defined by the length of the waves (*wavelength*) and the frequency of their passing. It is possible, therefore, just to list these energy types in a table with increasing wavelength. This is called the *electromagnetic spectrum* and the key divisions are shown in Figure 8.1.

The most important division in this spectrum is between *ionising* and *non-ionising* radiation. Essentially the difference between these is that ionising radiation can produce chemical changes as a result of ionising molecules on which it is incident. Non-ionising radiation, however, does not have this effect and is usually absorbed by the molecules on which it is incident, with the result that the material will

Figure 8.1: The Electromagnetic Spectrum

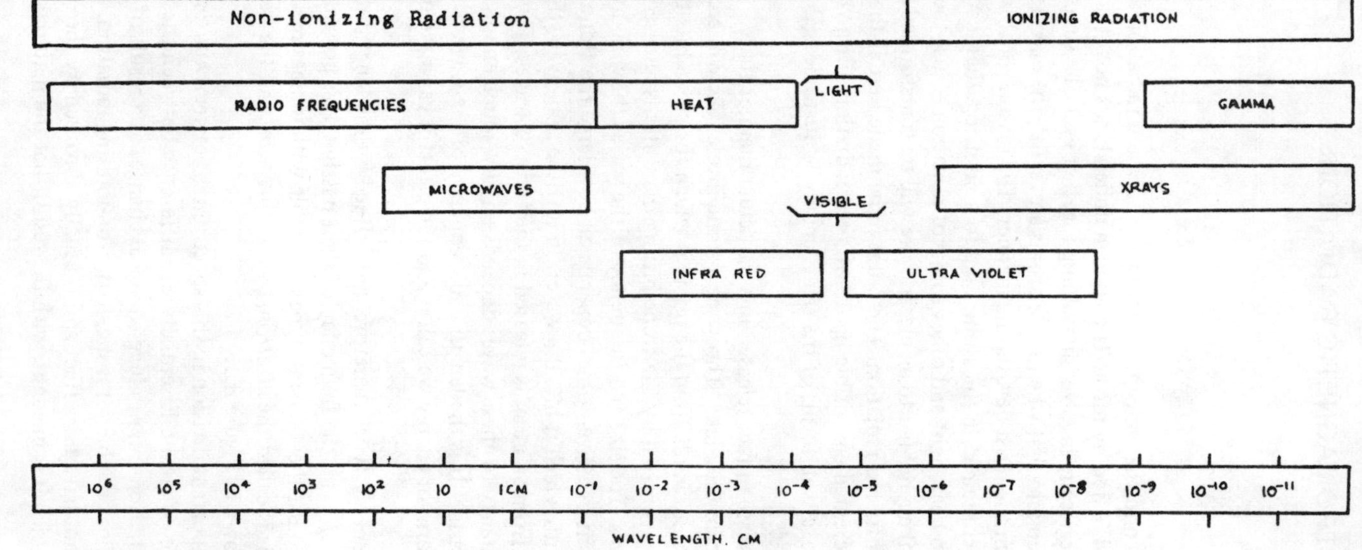

heat up, e.g. such as occurs with microwaves.

The purpose of this chapter will be to describe these two types of radiation, their effect on the human body, methods of their detection and guidelines for the limits of exposure.

Ionising Radiation

This range of radiation wavelengths can be sub-divided into two separate areas of interest, X-rays/gamma rays and ultra-violet radiation.

(1) X-Rays/Gamma Rays

These are considered *potentially* harmful no matter what the exposure level. By ionising the molecules in the living cell they can cause the formation of very reactive molecules called free radicals, with the result that large chemical changes can occur. The changes can result in the effects shown in Table 8.1.

Table 8.1: Potential Effects of X-Rays and Gamma Rays

System Affected	Effect
Blood and bone marrow	Anemias, leukemias, cancer
Lymphatic system	Spleen damage, lymphocytopenia
Skin	Erythema, degenerative changes, carcinomas, loss of hair
Eyes	Cataracts
Reproductive system	Sterility, genetic changes, lowered sperm or egg production
Lungs	Alveoli degeneration, cancer

Units. So that the degrees of effect from different radiations can all be related to the same unit of measurement, a unit called the *Rem* was conceived. To allow some perspective to be placed on radiation exposure Table 8.2 lists the average exposure per year of US individuals, in terms of each source of radiation (1970 data from US Environmental Protection Agency).

In terms of guidelines limits, the ACGIH TLV book accepts the recommendations of the US National Council on Radiation Protection and Measurements (NRCP). This also gives references to support the guidelines and these are listed at the end of this chapter.

The limit agreed, therefore, is that the dose to the whole body, when

Table 8.2: Average Radiation Exposure Per Year of US Citizens

Source of Exposure	Level (milliRems/year)
Natural (solar, cosmic, minerals)	130.0
Occupational (reactors, isotopes)	0.8
Nuclear power	0.002
Weapons fallout	4.0
Medical (X-rays – includes dental)	72.0

added to the *accumulated* occupational dose to the whole body, should not exceed 5(N-18) Rem, where N is the age of the individual in years at the last birthday, e.g. a man 35-years-old should not exceed 5(35-18) = 85 Rem. Any employee who is under 18 years of age, should not be permitted to receive in any period of one calendar quarter a dose in excess of 10 per cent of the limits set out in Table 8.3.

Table 8.3: Radiation Exposure Limits

System	Rems per Calendar Quarter
Whole body; head and trunk; blood-forming organs; eyes or gonads	1.25
Hand and forearms; feet and ankles	18.75
Skin of whole body	7.5

Measurement. The devices used to measure ionising radiation are all based upon the ability of this type of radiation to ionise, i.e. to create a charged body from neutral atoms. One of the commonest is the Geiger-Muller tube.

The Geiger-Muller survey instrument utilises a glass tube filled usually with argon gas which forms ions when radiation is incident upon it. Positive and negative electrodes exist at the ends of this tube and so the positive ions move to the negative electrode and the electrons (negative), formed during the formation of the positive ions, move to the positive electrode. The quenching action of these collisions with the electrodes can be measured electronically (since they cause a voltage pulse) thus giving a measure of the radiation strength.

Other forms of measuring device exist which can be used to measure actual worker dose. Film badges have been used extensively for this purpose and currently there is increasing use of thermoluminescent dosimeters. A comprehensive discussion of instrumentation can be found in *Radiation Protection Instrumentation and its Application,* published by the International Commission on Radiation Units and Measurements (1971).

Because of the potentially serious nature of exposure to this form of radiation, the assessment of exposure should be carried out by experts. Should an OH nurse or safety representative feel that there are sources of this form of radiation, then the wisest course of action would be to commission a consultant or discuss it with the supplier of the source.

(2) Ultra-violet (UV) Radiation

Less able to penetrate the body tissues, the effects of ultra-violet tend to be limited to the epithelium of the skin and to the eye structures. As with the shorter wavelength radiation ultra-violet can cause molecular dissociation and free-radical formation.

The term ultra-violet covers a range of wavelengths and the biological effects vary with this. For example, UV wavelengths of 250-300 nanometres (1 nm = 10^{-7} cm) can cause erythema of the skin, tanning, pigmentation and carcinogenesis. However, above a 320 nm wavelength these effects do not occur. Conjunctivitis occurs mainly in a narrow wavelength range around 270 nm. Other eye effects are regressive corneal lesions, photosensitivity and fluorescence.

Units. These are normally in power density or energy density, power density being given in Watts/cm^2 and energy density in Watt secs/cm^2 (or Joules/cm^2).

Limits. The ACGIH has proposed the following:

(1) For the UV region of 315-400 nm, total incident radiation on unprotected skin or eyes shall not exceed 1 m^J/cm^2 for periods greater than 1000 seconds and for exposure times less than 1000 seconds shall not exceed 1 J/cm^2.
(2) For UV in the region 200-315 nm, radiant exposure incident upon the unprotected skin or eyes should not exceed the values given in table 8.4 within an eight-hour period.

Measurement. Devices are available similar to light meters. Several are available but care must be taken in selection since not all have spectral characteristics that match those in the TLV. The International Light Inc. of Newburyport, Massachusetts, USA, developed one in 1974 in

Table 8.4: UV Exposure Limits

Wavelength (nm)	TLV (m J/cm^2)
200	100
210	40
220	25
230	16
240	10
250	7.0
254	6.0
260	4.6
270	3.0
280	3.4
290	4.7
300	10
305	50
310	200
315	1000

conjunction with NIOSH and this *does* match the TLV spectral range.

Control. Because UV is more common in industry than the lower-wavelength radiation it is as well to mention control measures. Darkened lens glasses are available for eye protection and are commonly used in welders' helmets. Skin can be protected by clothing or reflective creams and ointments.

Non-ionising Radiation

(1) Microwaves

As mentioned earlier these heat up molecules by increasing molecular rotation and kinetic energy so that the ultimate result is to heat objects. As with UV the actual biological effect depends on the wavelength as shown in Table 8.5.

Table 8.5: Biological Effects of Microwaves

Wavelength	Biological Effect
Less than 3 cm	*Some* heating of skin surface
3 cm	Skin heating with sensation of warmth
3-10 cm	Lens of eye can be affected, eye cataracts
25-200 cm	Internal organs can overheat
Greater than 200 cm	Pass straight through with no effect

Units. As for UV i.e. power density or energy density.

Limits. ACGIH's recommended TLVs state:

(1) For continuous exposure to continuous waves the power density should not exceed 10 mW/cm^2 over an exposure time of eight hours.
(2) Short-term exposures are permissible up to 25 mW/cm^2 based upon an average energy density of 1 milliwatt-hour per square centimetre, averaged over any 0.1 hour period. For example, at 25 mW/cm^2 the permissible exposure duration is approximately 2.4 minutes in any 0.1 hour (6 mins) period.

Further limits are available for pulsed microwave sources which the reader may find in the ACGIH book should the need arise.

Measurement. The devices available work by converting the microwave energy into heat and measuring the change in heat with a thermistor etc. Narda Microwave Corporation, New York, provides adequate instrumentation in this field.

Control. Probably the best means of control is to turn it off, i.e. to control the exposure time! Protective clothing and goggles are available, however, should it prove necessary to work in a microwave field, but otherwise the on/off switch is the best means of protection.

(2) Lasers

Lasers are man-made sources of radiation and they differ from the previous sources in that they operate at specific wavelengths and achieve great power densities. They operate mainly in the infra-red, ultra-violet and visible regions.

For this reason, the effects on the skin and eye can vary considerably. However, the primary effect is burning due to the high power density. This applies to both the skin and the eye. Wavelengths in the visible region, of course, are focused by the lens of the eye onto the retina. Lasers in this wavelength can therefore burn the retina very quickly when looked at directly. A particularly sensitive area of the retina is the fovea which contains the highest concentration of cone cells and damage to this region can result in serious loss of vision.

Units. As for UV and microwaves.

Limits. The ACGIH gives extensive data on the allowable exposure time for particular laser power densities and wavelengths. Also covered are direct viewing exposures and diffuse exposures. These standards are complex, but simple to use as long as the investigator has the laser characteristics available. In the UK, the 1974 Eye Protection Regulations also stipulate the use of specified goggles for laser work.

Control. In this case, non-reflecting shrouds should be used. The incident surface should be held secure and non-reflecting. Direct viewing should be made impossible unless the allowable exposure time is found to be several seconds at least. Ensure that the area is fully illuminated to prevent pupil dilation as this increases the possibility of penetration to the retina. Note that BS (British Standards) 4803 provides a great deal of information on the safe use of lasers.

(3) Infra-red

Infra-red penetrates more easily than ultra-violet particularly in the near-visible range. Although present in a number of industrial operations e.g. metal industries, it is primarily an eye hazard and as such is easily protected against by filtered goggles. The ACGIH limits it to 10 mW/cm^2 to avoid possible delayed effects upon the lens of the eye (cataractogenesis). The limit is based on a 7 mm pupil diameter.

Summary of Key Points

(1) The electromagnetic spectrum can be divided into ionising and non-ionising radiation.
(2) Ionising radiation can cause chemical changes in the living cell by production of free radicals.
(3) Non-ionising radiation causes molecules to increase in kinetic and rotational energy. As a result, the material heats up.
(4) Short wavelength ionising radiations such as X-rays and gamma rays are potentially dangerous at any level of exposure (carcinogenesis).
(5) Ultra-violet radiation can cause serious skin and eye problems but little penetrates the deep body tissues.
(6) Infra-red radiation can cause cataracts but is easily protected against.
(7) Microwaves can cause external skin damage and also the internal organs to heat up.

(8) Lasers usually have narrow wavelength ranges and high power densities. They can therefore cause severe burning. Particularly dangerous to the retina.
(9) ACGIH TLVs provide guidelines on exposure to all of the electromagnetic radiations.
(10) Wide range of measuring tools available but advice usually needed.

Further Reading

Basic Radiation Protection Criteria, Report 39 (National Council on Radiation Protection and Measurements, Washington, DC, 15 January 1971)

Manual of Respiratory Protection Against Airborne Radioactive Materials (US Atomic Energy Commission, Washington, DC, 1974)

Exposure to Radiation in an Emergency, Report 29 (National Council on Radiation Protection and Measurements, Washington, DC, January 1962)

Sowby, F.D. (ed.) *General Principles of Monitoring for Radiation Protection of Workers,* Publication 12 (Adopted by the International Commission on Radiological Protection, Sutton, Surrey, 24 May 1968)

Inhalation Risks from Radioactive Contaminants, Technical Report Series No. 142 (International Atomic Energy Agency, Vienna, 1973)

Klein, H.G. (ed.) *Radiation Safety and Protection in Industrial Applications,* Proceedings of a Symposium held 16-18 August 1972, Washington, DC (US Department of Health, Education and Welfare, Bureau of Radiological Health, Rockville, Maryland, October 1972)

Maximum Permissible Body Burdens and Maximum Permissible Concentrations of Radionuclides in Air and in Water for Occupational Exposure, Report 22 (National Council on Radiation Protection and Measurements, National Bureau of Standards, Handbook 69, 1959)

Permissible Dose From External Sources of Ionising Radiation, Report 17 (National Council on Radiation Protection and Measurements, National Bureau of Standards, Handbook 59, 1954; includes Addendum to NBS Handbook 59, issued April 1958)

Radiation Protection Instrumentation and Its Application, Report 20 (International Commission on Radiation Units and Measurements, Washington DC, 1 October 1971)

Safe Handling of Radioactive Materials, Report 30 (National Council on Radiation Protection and Measurements, National Bureau of Standards, Handbook 92, 1964)

9 INDUSTRIAL VENTILATION

Introduction

Occupational hygiene is concerned with the *recognition, evaluation* and *control* of health hazards in the workplace. Satisfactory control of a contaminant level is the biggest headache of the hygienist. Often, there are substantial reasons why a process unit or a piece of equipment cannot be prevented from releasing a contaminant into the workroom air. In this event, control of the hazard can only be achieved by either providing the workers with personal protection, ranging from barrier creams to full suits, or by removing the contaminant from the air so that it never reaches the work-force. This latter solution is the primary objective of industrial ventilation.

This chapter will concentrate on developing the reader's ability both to be constructively critical of an *existing* ventilation system and also to provide guidelines to plant engineers for the key design points to be incorporated into a *new* system. It is not the intention to provide information to allow the reader to carry out a detailed design, since this would require a separate text on its own. However, readers should regard their responsibility as defining the design specifications, i.e. setting out the task to be achieved by the system, including giving advice on how this can best be carried out. There will be no discussion of air conditioning systems since this is a subject related mainly to the 'comfort' of the operators and is outside the brief of this book.

Basically there are four areas of interest in industrial ventilation. First there are general rules regarding the *siting* of ductwork, air filters, exhaust ducting, etc. Secondly and thirdly there are the two main types of ventilation commonly used, ie. *dilution* and *local.* Fourthly there is the *testing* of these systems.

The principles and pitfalls of dilution and local ventilation will be discussed first, mainly in the form of examples to allow the reader to gain experience. Then the general rules on the overall system will be discussed – again by example – and finally, a section will describe the tools and techniques used in the testing of these systems.

For those who wish to extend their knowledge of this field there exists a very readable book that can be used as a reference text for future problems. Published by the ACGIH, it is republished every two years to allow new concepts to be included. Titled *Industrial*

Ventilation – A Manual of Recommended Practice, it is widely used by hygienists and ventilation engineers alike.

Dilution Ventilation

At the outset, we should distinguish clearly between dilution and local ventilation. As the name implies, dilution refers to the removal of general room air (containing the contaminant) and replacing or diluting this with 'fresh' air. In this way, the general level of contaminant in the room is kept down to a predefined, acceptable level. Local ventilation, again as the name implies, refers to the local removal of air from the area at which the contaminant is being released. In this case the contaminant, theoretically, is removed before it reaches the overall room environment.

In general, dilution ventilation is not as reliable as local ventilation for hazard control purposes. This is mainly because the contaminant is rarely distributed evenly throughout the room. It is usually only recommended, if the following situations exist.

(1) The contaminant is released at a *low rate* and not from a point source.
(2) The contaminant is released at a *uniform rate.*
(3) The contaminant is of *low toxicity.*
(4) The worker is not employed particularly close to the source of release.

Note that measurement of the rate of release can be achieved by any of the methods described in the individual chapters in this book. If any doubts exist on any of the four above points, then dilution ventilation will not be an effective measure for controlling the health hazard. However, assuming that the above situations do exist, as they can do in laboratories where low levels of innocuous solvents may be released slowly, then the rate of air change necessary must be assessed.

Suppose the room is 20 m by 10 m by 3 m, i.e. a volume of 600 m^3. The contaminant is say mainly n-heptane with a TLV of 400 ppm. Obviously, we would wish to maintain a room level much lower than this, so take 40 ppm (normally a factor of 10 as a safety margin is a good baseline position). By measurement or calculation, the rate at which the solvent is released into the air is found to be 4,000 cm^3 of vapour per minute (from all sources).

In one hour, therefore, the volume of vapour will have reached 240,000 cm^3. If the ventilation changes the room air N times in one hour, then the ppm level of vapour will be:

$$\left(\frac{240{,}000}{N \times 600}\right) \text{ ppm}$$ (N.B. ppm by volume is the same as cm^3 of vapour/m^3 of air.)

But we wish the limit to be 40 ppm. Therefore, let

$$\left(\frac{240{,}000}{N \times 600}\right) = 40 \text{ i.e. } N = 10$$

This then tells us that we need ten room air-changes per hour under these conditions to maintain the 40 ppm limit in this room. The volume of air to be moved by the fan is therefore:

$$\begin{aligned} 10 \times 600 &= 6{,}000 \text{ m}^3\text{/hour} \\ &= 100 \text{ m}^3\text{/minute} \end{aligned}$$

Giving this information to the design engineer will allow him to select the right size fan. There is also some more information that the investigator can give the engineer: where to site the fan and where to site the air inlet.

To show the importance of this we can consider a simple dilution ventilation problem as follows. Jim Stickman works at a bench cleaning motor coils using a commercial solvent which he puts on a cleaning cloth from a topped drum on the bench. Low levels of solvent are slowly released during this cleaning. We know from the suppliers of this solvent that it is relatively innocuous with a TLV of 500 ppm and we have rightly selected dilution ventilation and have sized the fan. The engineer now wants to know where to put it.

Consider the orientation of the room as shown in Figure 9.1 (S being the source of the contaminant).

The first thought is to site the fan on the back wall away from the man with an exit alongside (Figure 9.2).

The problem here is that the inlet air will be contaminated with the air leaving the room so that the level of contaminant will build up.

An alternative therefore, is to move the air inlet to the other side (Figure 9.3).

This solves the first problem but as the reader will see, the fresh air

Figure 9.1

Figure 9.2

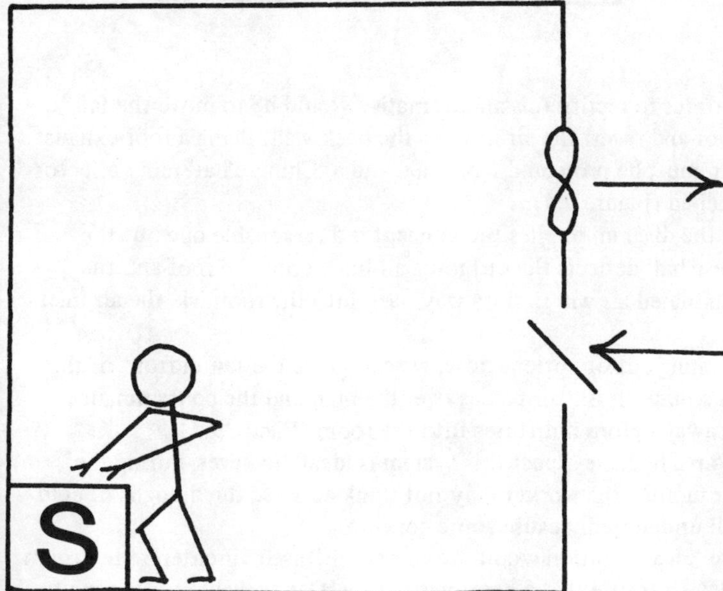

moves across the 'source', past the operator and then out. The result is that the contaminant is blown straight at the operator and the situation is made worse.

Figure 9.3

In order to rectify this an alternative would be to move the fan to the roof and revert the air inlet to the back wall. Being a roof exhaust the fan must be protected from rain and a 'Chinese hat' rain deflector is attached (Figure 9.4).

As the diagram implies the concept is a reasonable one but the 'Chinese hat' deflects the outgoing air back onto the roof and the contaminated air will find its way back into the room via the air inlet again.

Running out of options now, we can move the fan in front of the bench so that 'fresh' air passes over the man and the contaminant is taken away before it diffuses into the room (Figure 9.5).

From a hygiene aspect this system is ideal; however, during the winter months the worker may not think so since the draught of cold air will undoubtedly cause some concern.

The 'ideal' solution would be to have diffused air entering the room and also to ensure that a heating system (H) is available for the winter months (Figure 9.6).

Figure 9.4

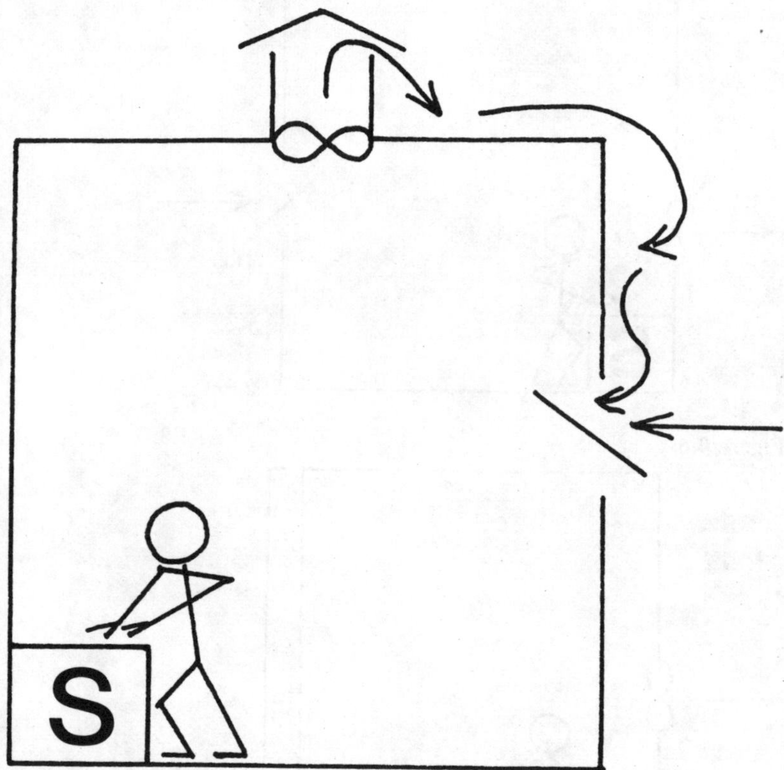

This example serves to show that with dilution ventilation the siting of the fan and air inlet are critical. The rules normally to be applied can be summarised as:

(1) Site the exhaust fan near to the source of the contaminant.
(2) Ensure that fresh air movement is from man to source and not vice versa.
(3) Ensure that the inlet air supply is not contaminated with the exhaust air.
(4) Provide back-up heating for the inlet air for cold climatic conditions.

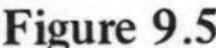

Figure 9.5

Figure 9.6

Local Ventilation

The more effective system is local ventilation, but the design is much more critical and quite often seriously underestimated. The principle is to capture the contaminant in an airstream *at the point of emission* – or near enough to ensure capture.

Dilution ventilation is usually only acceptable if the contaminant is a gas or vapour since fumes and dust are not normally so evenly distributed within a work space. Local ventilation can be used for all of these contaminants.

In order to achieve capture there are four basic strategies:

(1) Completely *enclose* the source and continuously remove contaminated air from the enclosure.
(2) Place a *hood* over the process to draw the contaminant *away* from the operator.
(3) Position an extraction *slot* at the side of a process to draw the contaminant away from the operator.
(4) Place an extracted *container* in the path of pollutants which have a velocity as a result of the process, e.g. from grinding, sawing, etc.

Before entering the more complex world of designing for the correct air flow rates, it is as well to ponder the potential pitfalls of simply locating these local ventilation systems. Explanations are probably not necessary, but suffice to say that:

(1) Hoods should not be used if the man needs to lean over the process to do his work.
(2) Where hoods are used ensure that they cover the point of release of the contaminant. If powder is falling into a hopper from a conveyor, the major dust emission will be from the base of the hopper. The hood will, therefore, need to reach *down to the hopper* to ensure collection of the dust.
(3) If a container device is used, e.g. extraction hood in the path of a pollutant moving at high velocity, then *ensure that the path is covered by the hood.*

The next most usual mistake in the design of ventilation systems is to underestimate the amount of air necessary to 'capture' the pollutant.

The amount of air necessary will in fact depend on the type of the pollutant. The critical variable here is the velocity of the air necessary to capture the material and take it into the control system – this is known as the *capture velocity.* The ACGIH manual gives a listing of approximate velocities that are necessary to capture various contaminants and these are reproduced in Table 9.1.

Of course, the capture velocity is the velocity required at the contaminant point itself which may be several centimetres away from the hood. The velocity across the face of the hood is known as the *face velocity.* This velocity is the one that can be fixed by design, i.e. by the volume of air flow rate and the cross-sectional area of the hood. Thus if a hood is 1.2 m x 60 cm and a face velocity of 30 m s^{-1} is required, then the volume flow rate of air needed is:

$$1.2 \times 0.6 \times 30 = 21.6 \text{ m}^3 \text{ s}^{-1}$$

Figure 9.7 shows some typical blunders in locating hoods, slots and containers.

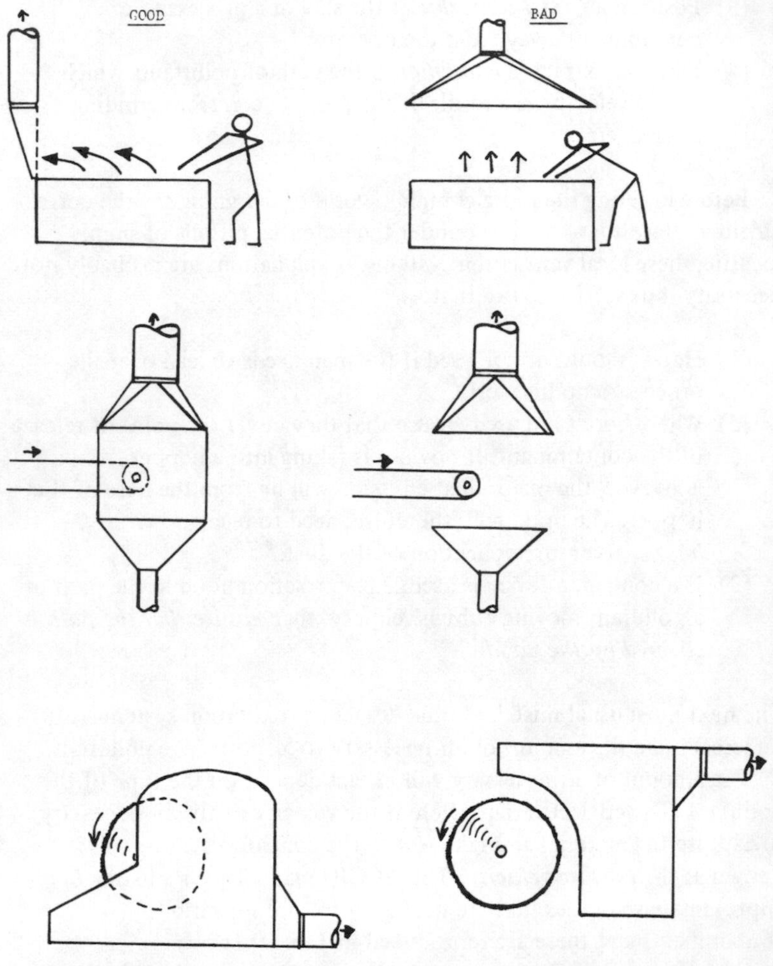

It is beyond the scope of this book to go into the detail of assessing air velocities in front of hoods/slots. Suffice it to say that nomograms have been produced by Fletcher (see *Annals of British Occupational Hygiene Society*, vol. 20, no.2 (October 1977) and vol. 21, no.3 (December 1978)) which allow prediction of the velocities at distances

Table 9.1: Velocities Needed to Capture Various Contaminants

Condition of Contaminant	Examples	Capture Velocity (m s^{-1})
Released with practically no velocity in still air	Evaporation from tanks, degreasing, etc.	0.25-0.5
Released at low velocity in moderately still air	Spray booths, low-speed conveyors, welding, plating	0.5-1
Active generation into zone of rapid air motion	Crushers, conveyor loading	1-2.5
Released at high initial velocity into zone of very rapid air motion	Grinding, blasting	2.5-10

from hoods and slots if the dimensions of the hood and the face velocity are known.

A point of note here, however, is that when slots are used with an aspect ratio width/length of 0.2 or less, the efficiency of the system can be increased by adding a flange to the slot – usually with a flange width at least equal to the slot width. This reduces the amount of air drawn in from *behind* the slot (not contaminated) so that more air is taken from in front (where it is needed). As a result, the capture velocity is increased at any particular distance in front of the hood (or the quantity of air could be reduced for energy conservation if the hood is already correctly positioned).

To allow the investigator to gauge the loss of velocity that occurs in front of slots and hoods the following example may prove both helpful and surprising. If we have a slot width of 5 cm, a slot length of 40 cm and a face velocity of 20 m s^{-1}, then using Fletcher's nomograms, the velocities at different distances in front of the slot can be calculated as shown in Table 9.2.

Table 9.2: Velocities Calculated Using Fletcher's Nomograms

Distance in Front of Slot (cm)	Air Velocity (m s^{-1}) Produced by	
	Unflanged Slot	Flanged Slot
0 (face velocity)	20	20
5	4.6	6.5
10	1.6	2.5
15	0.8	1.3
20	0.5	0.8
30	0.25	0.4
40	0.15	0.25

The reader will note that in *both* cases the velocity has dropped significantly even only 5 cm (c.2 inches) away from the slot. Hopefully this will give readers the correct impression that slots placed even only one metre away from the source will probably be ineffective.

Other guidelines for improving the overall capture efficiencies are:

(1) Take advantage of natural currents of air and pollutants, i.e. hotter air will rise, so use an *overhead* hood for hot processes.
(2) Slots of over two metres length should have more than one offtake for the air.
(3) A 'fish-tail' plenum (enclosed space behind the slot or hood) should be used to ensure good distribution across the face.
(4) Enclose as much of the unit as is physically possible without hindering the worker.
(5) When wide areas are to be locally ventilated, it can be necessary to use a *supply* air slot to *push* the contaminants towards the exhaust slot, since the 'throw' of a jet is much longer than the pull (see Figure 9.8).

Figure 9.8: Supply Air Slot

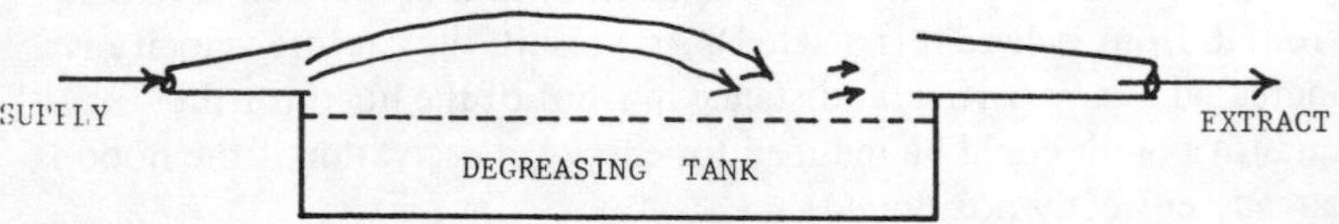

General Rules on Design

Having achieved an ideal design and siting of the extract inlet, several other aspects of the system require some explanation.

(1) Transport Ducting

Having collected the contaminant the ductwork must be able to transport it out of harm's way. This is particularly critical if the contaminant is particulate, i.e. a dust. The velocity in the ductwork must be sufficient to maintain the material airborne, otherwise ducts will become blocked and the system will fail. The velocity that can achieve this is known as the *transport velocity*. The ACGIH manual again gives guidelines as to the transport velocities that may be required as shown in Table 9.3.

Table 9.3: Transport Velocities for Various Contaminants

Nature of Contaminant	Examples	Transport Velocity (m s^{-1})
Vapours, gases, smoke	Obvious	Any, but economics usually indicate 5-6
Fumes	Welding	7-10
Very fine light dust	Cotton, wood, flour	10-13
Dry dust and powders	Soap powder, jute dust, rubber dust	13-18
Average industrial dust	Granite, silica, limestone etc.	18-20
Heavy dusts	Metal turnings, sand blast, lead dust	20-23
Heavy or moist dusts	Moist cement dust, lead dust with chips	>23

As an example, suppose an average industrial dust is being extracted through a slot the design of which necessitates an air flow rate of 10 m^3 s^{-1}. From Table 9.3, we can work out that the transport velocity should be around 20 m s^{-1}, since

Volume flow = Area of duct x Velocity of air
therefore 10 = Area of duct x 20
therefore Area of Duct = 0.5 m^2
i.e. a duct diameter of 0.8 metres or 80 cm (assuming a round duct is used)

Therefore, an 80 cm diameter duct will be required to maintain the required transport velocity.

(2) Duct Orientation

For air to be able to move within a duct or pipe it must overcome the friction between itself and the walls of the duct. This requires energy which is dissipated or lost from the pressure of the air. Air movement therefore can only be achieved with a resulting pressure loss or drop in the direction of the movement.

In order to reduce the pressure loss in the system (and therefore cost) and also to avoid areas where the air impacts the ducting (causing particles of dust to adhere), ductwork should have 'smooth' junctions. This means avoiding right-angle bends, square junctions, etc. shown in Figure 9.9.

Figure 9.9: Design of Junctions to Minimise Pressure Loss

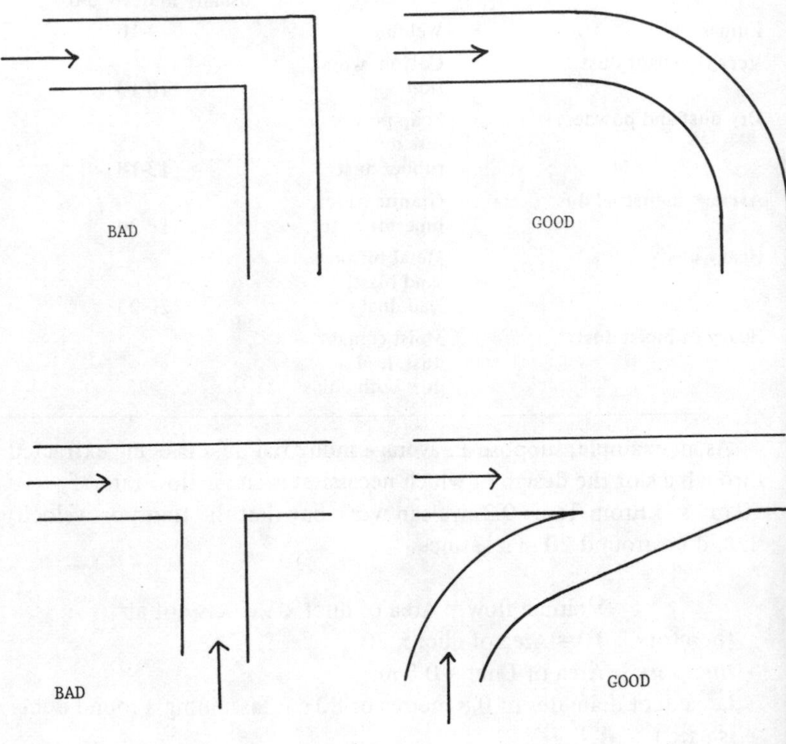

(3) Positive Pressure Outside the Building

Although the ducting before the fan will be at negative pressure (therefore any leaks cause room air to go *into* the duct) the ducting *after* the fan will be under positive pressure. Any leaks in the positive pressure side will cause the contaminated air to *leave* the duct. All positive-pressure ducting should therefore be outside. The best way to guarantee this is to actually have the fan itself located outside.

(4) Chinese Hat?

As mentioned earlier, Chinese-hat rain protectors cause the exhaust air to be deflected back down towards the building. Since other, equally efficient but non-deflecting rain traps exist, the Chinese hat should never be recommended. The Chinese hat also causes a high pressure loss and increases the running costs of the system. Many alternatives exist which do not deflect downwards and reference to a textbook on ventilation will show these.

(5) Filters

It is sometimes required that the extracted air be cleaned before sending it in to the atmosphere. Since a vast range of filters exists and also since a degree of specialist knowledge is required these will not be discussed here. Suffice to say that the ACGIH manual covers these very well and reference can be made to this when necessary.

(6) Blast Gates (Simple Gate-type Dampers to Restrict Air Flow)

A number of designers still allow flexibility in the use of ventilation systems by providing *blast gates* in the ducting which allow an operator to alter the flow of air to suit particular requirements. If flexibility is essential (it should not be if the system is designed correctly), then the hygiene investigator should ensure that a regular check is made on the condition of the gates *and* to determine if the operators are using them properly – if at all.

Testing of Ventilation Systems

For full-scale testing of ventilation systems, complex procedures and testing tools are required. We shall limit the discussion here to measuring the collection efficiency of the hood/slot. Should the efficiency be impaired then the investigator should inform the plant engineer who can then pinpoint the cause – after checking that the original design criteria were adequate.

The only quantitative measurement necessary, therefore, is *air velocity*. Several devices exist for making this, but three are of particular interest.

(1) Swinging-vane Anemometer (Velometer)

This instrument has an internal vane which swings about on its axis. The air is directed through a tube and deflects the vane. The degree of deflection is then read directly on the meter. Because of this, the meter should always be read when in an upright position. This is a major disadvantage since exhaust systems are not always located near a horizontal surface on which to rest the meter. As with all air velocity meters, calibration should be carried out regularly, but with this meter it is particularly important since the length of the connecting tube can be varied and the calibration will require repeating with each different length of tube.

(2) Heated-wire Anemometer

The action of this device depends on the variation of electrical resistance of a wire with temperature. When air moves over the heated wire, the amount of heat removed from the wire depends on the air velocity which can be measured directly. Calibration is again regularly needed, in this case because the wire can become coated or corroded and this will change the electrical resistance.

The Thermister Bead Anemometer operates under the same principle but has fewer problems with corrosion, etc. This utilises a semi-conductor of which the resistance is extremely high and which can give large changes with small cooling effects thereby giving good sensitivity (see Figure 6.1, page 92).

(3) Heated Thermocouple Anemometer

The same principle is involved here except that a thermocouple junction is used to detect the change in temperature of a heated probe in the air stream. A second thermocouple junction senses the temperature of the air so that the reading is independent of temperature. The probes tend to be fragile and need handling with care. Frequent calibration is necessary. These are either high-reading or low-reading instruments – not usually accurate at *both* ends of the scale.

The comparative characteristics of these three types of instrument are shown in Table 9.4.

Regardless of the type used the measurement of velocities at hood entrances and beyond, should be treated scientifically. One reading is not enough. The investigator should traverse the face taking several readings at different points (at least six readings for a normal slot). This is even more important in the free space in front of the hood/slot as external winds or draughts may cause large fluctuations in the extractor efficiency.

Other air velocity devices include:

(1) Pitot tube (used for duct velocity measurements).

(2) Rotating-vane anemometer (similar to wind cups but much more sensitive – used mainly for open ends of ducts).
(3) Orifice meter (used for duct measurements and needs pressure difference measurements to be taken).

Note that calibration of these instruments should be done in an open jet wind tunnel which can be found in most universities.

Table 9.4: Characteristics of Instruments Used to Measure Air Velocity

Type	Range of Air Velocity ($m\ s^{-1}$)	Dust, Fume Difficulty	Ruggedness	General Comments
Swinging vane	0.1-50	Can be a problem due to internal vane collecting dust	Only fair	Useful for 'easy' measurements
Heated wire (heated bead)	0.05-40	Care must be taken – short life if too much dust in air (bead version is better)	Good	Very portable, easy to use
Heated thermocouple	0.05-10 or 2.5-50	Can be severe problem	Poor	Good if working conditions are easy

A final tool that every investigator should have is a box of *smoke tubes.* These are glass vials whose ends are snapped off and a rubber squeeze ball is used to blow air through the tubes. Smoke is produced (by various means depending on the tube) and can be used to gain a visual assessment of the effectiveness of the hood/slot without resorting to the less controllable smoke pellets or bombs.

As a concluding message to this chapter the reader should remember that the measurement of the level of contaminant in the environment is always a good measure of ventilation effectiveness since that is the job for which the system is intended.

Summary of Key Points

(1) *Industrial Ventilation* (IV) is one of the main methods of control of the workroom environment.
(2) IV can be achieved by either *dilution* or *local* extract systems.

(3) *Dilution* is only advisable for low-toxicity gas or vapour contaminants and only when they are released at low, uniform rates.
(4) *Siting* of the extract and inlet is critical to ensure an efficient dilution system.
(5) *Local* ventilation must be carefully designed to ensure that the *correct capture velocity* is achieved at the point at which the contaminant is emitted.
(6) The ACGIH manual gives guidance on capture and transport velocities.
(7) *Flanging* a slot improves the efficiency of capture.
(8) *Hood air velocity* measurements can be carried out using:

(a) swinging-vane anemometer (velometer);
(b) heated-wire (or bead) anemometer;
(c) heated thermocouple anemometer.

(9) *Calibration* of air velocity meters should be done regularly in an *open jet wind tunnel.*

Further Reading

ACGIH *Industrial Ventilation – A Manual of Recommended Practise*, 15th edn (ACGIH, Ann Arbor, Michigan, 1978)

10 PERSONAL PROTECTIVE EQUIPMENT

Introduction

As the reader will guess, the range of protective equipment is vast. The use of this type of equipment is mainly to allow tasks to be completed which would otherwise be potentially dangerous to health. For this reason, it should normally be regarded as the last line of defence. Only when it has proven impractical to control the hazard in other ways should this line of action be taken. One of the key problems with this strategy is that it places a great deal of emphasis on the user. With this in mind we have included a section dealing with the *management* of personal protection programmes.

It is not intended to cover in detail the individual areas of equipment except respiratory protection. The other areas will receive a cursory examination, but the reader is provided with a list of specialist references for each of these, together with the extensive legislation on individual processes to indicate the legal concern for this important topic. Obviously the range of equipment available is vast and cannot be covered in detail here but sufficient references are given so that further information can be gathered when a protection programme is being set up.

(1) Eye Protection

This is of great importance and is one of the few aspects where worker participation is assured. Difficulties of credibility exist when hygienists attempt to convince operators of the need to wear ear protection, dust-masks, gloves or aprons – but this is almost never the case with eye protection. The risk factor is all too obvious and the need to wear protection rarely requires a second telling. From a legal standpoint, this is also an important concern. The 1974 'Protection of Eyes' regulations stipulate the processes that require eye protection and in what form. Only British Standard (BS) approved protection should be provided for these processes, the most relevant being BS 2092. Schedules I and II of these regulations are appended to this chapter for reference.

As with all forms of protective equipment, the most important selection criterion is to ensure that – in this case – the goggles and shields bear the British Standard kitemark. The suppliers will also have the information as to the section of use to which the kitemark refers. Always remember that the design of the eye protection must match the task it has to perform, e.g. against dusts, gases or liquids.

(2) Protection of Skin and Body

Several aspects are of interest in this category and it is important that the investigator quickly decides which is the area of concern.

(a) Prevention of Contamination

This refers to the situation when contact with the work materials is not injurious to health but merely undesirable in terms of the worker exposing himself or his normal clothing. For example, a mechanic would not wish to work on a car in his normal shirt and trousers. Clothes worn in these situations are usually not completely impervious to liquids or dusts and the control involved in ensuring the wearing and changing of these garments is less stringent than in other, more hazardous situations.

The responsibility of the employer in these cases is to ensure that good changing facilities are available and that the clothes are regularly cleaned.

(b) Protection against Corrosive Chemicals

Where a material can have an immediate damaging effect on the skin (e.g. acids), the clothing must be impervious to the contaminant. Usually it will be impregnated with rubber or plastic to achieve this. The problem here is that impervious clothing also prevents sweat evaporation. Care must be taken, therefore, to ensure that the worker does not suffer from heat strain as a result of working too hard for too long when protection such as this is necessary. Air-supplied suits exist which provide 'fresh' air to the local space around the body should the need arise. The design of these suits is as yet not optimum and their use must be closely watched to ensure that the correct air rate is supplied and that good distribution is achieved inside the suit.

(c) Protection of Skin against Toxic Materials

Some chemicals will penetrate the skin and produce systemic and/or local toxic effects. Examples are hydrogen cyanide and some polynuclear aromatics (as can be found in cutting oils). A good example of an actual situation is the machine operator working with cutting oils. When an operator wears a cotton apron, for example, the oil builds up, particularly around the groin area. Unless frequent washing and changing of the apron occurs, the apron can become saturated with oil which can easily penetrate through to the skin. Cases of cancer of the scrotum have arisen due to this cause.

The solution is similar to that of the previous category except that extra care must be taken by management to ensure that the protection is worn and changed regularly. In the case of the machinist, special rubber/plastic-backed aprons now exist so that absolutely no oil can penetrate even if the aprons are not changed as frequently as they should be. These also have an absorbent, detachable front which precludes oil from running down the apron onto the lower limbs and which can be removed and sent for cleaning.

(3) Protection of Hands and Feet

Little needs to be said here except that a wide variety of gloves and shoes exists depending on the hazard that needs to be protected against. A good supplier will provide all of the necessary information on the applicability of the materials of construction with the hazardous chemical against which protection is required.

(4) Respiratory Protection

The lung is an external organ (see Chapter 1). Because of its nature, it is directly exposed to contaminants in the atmosphere. It also provides the quickest route for penetration into the bloodstream and thence to the main organs. Since the alveoli are the transfer media for the gas exchange necessary for existence, they are also transfer media for other airborne contaminants. Physiological problems resulting from inhaling contaminants can arise due to a number of causes as discussed in Chapter 3, but the importance of the efficiency of protection cannot be over-emphasised.

Naturally, the degree of protection depends on the toxicity of the material and the duration of the exposure. The type of protective device depends on the physical form of the contaminant. A wide range of protective devices exists providing breathable air in environments that vary from immediately dangerous to life to those containing relatively harmless, nuisance dusts. Consideration will be given here to the various forms that exist depending on the physical form of the contaminant.

Before discussing these in more depth, it is worth mentioning the degree of effectiveness of these devices – no matter what contaminant is being protected against. Obviously, this is vital information to the investigator if the right type is to be selected. This effectiveness is best described by what is called Nominal Protection Factor (NPF). NPF relates the ratio of the contaminant level in the atmosphere to the level in the air reaching the lungs. For example, a mask with an NPF of 10 means that if the atmosphere contains 50 mg/m^3 of dust then, with the mask, the wearer will breath air containing about 5 mg/m^3. This is important because the investigator may say that the level that is regarded as acceptable is perhaps 10 mg/m^3 but the environment contains 150 mg/m^3; therefore, in this case, a mask is needed with an NPF of 15. The key factors that control the NPF are: (1) the 'fit' of the rubber face seal to avoid leaks, and (2) the efficiency of the contaminant filter.

Because of the importance of the NPF, British Standards lay down procedures for its measurement and BS 4274: 1974 provides recommendations for the selection, use and maintenance of respiratory protection equipment. For dust collection, the standard half-face mask with screw-on filter will have an NPF of 7-10 whereas positive-pressure dust respirators (air fed by a pump through the filter) will have an NPF of 20-500 depending on the seal, etc.

(a) Dust Masks

Atmospheres containing dusts are cleaned by the use of filters made of materials such as paper, wool and fibre glass. The particles of dust are removed by several mechanisms, i.e. sieving, impaction, electrostatic forces. In all cases, however, the dust will obviously build up and as this occurs, the pressure drop across the filter will increase and the wearer will have more difficulty in breathing (see Figure 10.1). This will depend on the type of filter used; thus felt will take large dust loads with only small increases in pressure difference, whereas with cotton the pressure difference increases rapidly with dust load. Primarily, this is due to the fact that felt is an interwoven complex fabric of which the passages are numerous and require large numbers of particles to block them, whereas cotton fabric has only a small number of passages available for air movement.

Filters are tested under British Standard 2091 and all masks and

Figure 10.1: Typical Negative-pressure Respirators ('Dust Masks'). The one on the left covers the nose and mouth (ori-nasal) while the other is described as a full-face type. Both respirators use changeable cartridges which are chosen according to the contaminant hazard risk (either dust, gas or vapour).

filters bought should be checked to ensure that they are kitemarked to this standard for the type of dust concerned. Type A dust is low toxicity and type B is higher resistance for finer dusts. Note that asbestos dust filters are covered separately under BS 4275 and likewise cotton dusts under BS 4555.

(b) Gas Masks (Vapours also)

These work on a completely different principle to that of dust masks and should not be used for dust environments. The gas or vapour is removed by the chemical or physical adsorption of the molecules onto a solid material such as activated charcoal, silica gel or activated alumina.

The 'filters' of adsorbent material used in these masks have a much more well-defined lifetime than dust masks. Once the material has a particular gas or vapour load, it will then let the contaminant through to the wearer. Higher standards of contaminant removal and 'seal' are necessary in gas or vapour environments and types fall into two categories: *canister* type for specific toxic gases and *cartridge* type for relatively less toxic gases. Again the British Standards have recommendations for the design of these including the general BS 4275, but also ones for specific gases/vapours: hydrogen cyanide – BS 2091; ammonia – BS 4275; diethyl ether – BS 2091. Prohibition of the use of some masks exists where the TLV is less than 100 ppm.

The key point to remember here is that the canister or cartridge will only last a certain length of time and the system for replacing these must be strictly controlled and adhered to.

(c) Types other than Face Masks

As mentioned earlier, it is not necessary to rely on filtering media to remove the contaminant from the air being breathed. Alternatively, the worker may wear a hood or full suit to which air is supplied (see Figure 10.2). In this situation, the wearer may be subjected to extreme discomfort both thermally and physically. For this reason, the use of these should be limited and only under constant supervision. This supervision is necessary in any event due to the need to provide assistance should the air supply fail.

(5) Respiratory Protection Programmes

The use of a respiratory protection device implies that the worker is expected to operate in a potentially hazardous environment with the protection device as the key means of preventing exposure. It is, therefore, quite inadequate merely to provide the equipment and withdraw all further interest in the affair. A full respiratory protection programme must be set up. There are three key elements involved in this:

(1) the administration of the programme;
(2) the equipment involved;
(3) the training of the personnel.

These three facets should be regarded as vital to the success of the

Figure 10.2: 'Air-stream' Helmet. This uses a pump (positive pressure) to force air through a filter in the roof of the helmet and thence the clean air is blown over the face of the wearer. The face visor also gives eye protection.

programme since the failure of one will result in failure of the whole. Designation of responsibility is therefore important. Although an occupational health nurse would be capable of carrying out the programme, it is better if line management take this responsibility in order to gain involvement.

(a) Administration

There are three key aspects involved in the administration:

(i) Develop and obtain agreement to a proposed programme. Implicit in this is the involvement of an expert such as an industrial hygienist.
(ii) Co-ordinate setting up the programme.
(iii) Supervise the implementation of the programme.

(b) Equipment Selection, Use and Maintenance

BS 4275 covers these in some detail and anyone involved in this should take note of this standard.

Selection means defining the hazard, the work regime and then the equipment. The 'use' means defining the written procedures required for handling, storing, renewing and wearing the respirator. No respirators should be shared between workers otherwise no one man will take responsibility. Maintenance involves inspection, cleaning and recharging. Again, these instructions should be written.

(c) Training

It is vital that the wearer is aware of the reason for the protective equipment and what the limitations are. A good checklist for questions and points to be raised with the worker is the following:

(1) What is the nature of the hazard and what happens to the worker if exposure occurs?
(2) Why is it not possible to provide an engineering solution to the hazard?
(3) Why was this particular equipment selected?
(4) What is the equipment capable of and what are its limitations?
(5) How is it operated?
(6) Both theoretical and practical instruction must be given – including what to do if the equipment fails.
(7) Demonstrate the equipment in actual hazardous situations.

The personnel involved with the issue and maintenance of equipment must also be trained.

If all three areas of *administration, equipment* and *training* are attended to then the programme will be successful. If just one fails, then the whole will fail and the worker will not be adequately protected. The member of the health team involved must therefore ensure that whosoever is given the responsibility for the programme, utilises the necessary expert help to ensure success.

(6) Legislation Relating to Protective Clothing and Equipment

(1) *Aerated Water Regulations 1921 (No. 1932).* Provision of face, neck and arm protection for persons filling bottles.

(2) *Blasting (Castings and Other Articles) Special Regulations 1949 (No. 2225).* Provision of protective helmets, gauntlets and overalls. Provision for storage and maintenance.
(3) *Factories (Bronzing in Printing and Lithographic Processes) Regns. 1912 (No. 361).* Provision of overalls, with suitable headcoverings for females, and for storage.
(4) *Construction (Health and Welfare) Regns. 1966 (No. 95).* Provision of adequate and suitable protective equipment for persons obliged to work in open air.
(5) *Cement Works Welfare Order 1930 (No. 94).* Provision of waterproof boots and coats, goggles and head coverings for females.
(6) *Chemical Works Regns. 1922 (No. 731).* Provision of overalls and gloves, and in some instances respiratory protection (chrome process) and protective footwear.
(7) *Chrome Plating Regns. 1931 (No. 455).* Provision of bibs, gloves and waterproof footwear of suitable material with drying and storage facilities.
(8) *Clay Works (Welfare) Special Regns. 1948 (No. 1547).* Provision of suitable protective clothing for outdoor workers with provision for cleaning and storage.
(9) *Dyeing (Use of Bichromate of Potassium or Sodium) Welfare Order 1918.*
(10) *Factories (East Indian Wool) Regns. 1908.*
(11) *Electric Accumulator Regulations 1925 (No. 28).*
(12) *Factories (Flax and Tow Spinning and Weaving) Regns. 1906 (No. 177).*
(13) *Fruit Preserving (Welfare) Order 1919 (No. 1136).*
(14) *Factories (Hollow Ware and Galvanising Welfare) Order 1921.*
(15) *Factories (Glass Bevelling Welfare) Order 1920.*
(16) *Gut Scraping and Tripe Dressing (Welfare) Order 1920.*
(17) *Factories (Horsehair Processes) Regns. 1907 (No. 984).*
(18) *Indiarubber Regns. 1922 (No. 829)*
(19) *Iron and Steel Foundries Regns. 1953 (No. 1464).*
(20) *Laundry Workers (Welfare) Order 1920.*
(21) *Lead Compounds Manufacturing Order 1921 (No. 1443).*
(22) *Lead Paint Manufacture Regns. 1907 (No. 17).*
(23) *Lead Paint Regns. 1927 (No. 847).*
(24) *Factories (Lead Smelting etc.) Regns. 1911 (No. 752).*
(25) *Magnesium Special Regns. 1946 (No. 2107).*
(26) *Pottery (Health and Welfare) Special Regns. 1950 (No. 65).*

(27) *Sacks (Cleaning and Repairing) Welfare Order 1927.*
(28) *Ship Building and Ship Repairing Regns. 1960 (No. 1932).*
(29) *Tanning (Two Bath Process) Welfare Order 1918.*
(30) *Tanning Welfare Order 1930.*
(31) *Factories (Tin or Terne Plates Manuf.) Order 1917.*
(32) *Vehicle Painting Regns. 1926 (No. 299).*
(33) *Vitreous Enamelling of Metal or Glass Regns. 1908 (No. 1258).*
(34) *Factories (Heading of Lead Compound Dyed Yarns) Regns. 1907 (No. 616).*
(35) *Asbestos Regulations 1969 (No. 690).*
(36) *Grinding of Metals (Miscellaneous Industries) Special Regns. 1925 and 1950.*
(37) *Non-ferrous Metals (Melting and Founding) Regns. 1962.*
(38) *Certificate of Approval (Resp. Protective Equipment) 1977 Form 2486.*
(39) *Factories Act 1961 Sec. 30 Confined Spaces. BA Fan Approved Type.*
(40) *Breathing Apparatus etc. (Report on Examination) Order 1961 SI 1345.*
(41) *Jute (Safety, Health and Welfare) Regns.*
(42) *Diving Operations Special Regns. 1960.*
(43) *Protection of Eyes Regns. 1974.*
(44) *Lead Smelting and Manufacture Regns. 1911.*

(7) Schedules I and II of Protection of Eyes Regulations 1974

Approved eye protectors required for:

Shot-blasting of buildings.
Use of high-pressure water jets.
Use of cartridge tools.
Use of hammer, chisel or punch.
Chipping of paints, slag or rust.
Use of metal cutting saws or abrasive discs.
Pouring or skimming of molten metal.
Operations at molten salt baths.
Work in the presence of acids, alkalis and dangerous corrosive materials.
Injection of liquids into building structures.
Breaking up, cutting or dressing of metal castings, glass,

plastics, plaster, slag, stone, etc.
Removal of swarf by compressed air.
Furnace and foundry work.
Wire rope manufacture.
Cutting of wire or strapping under tension.
Glass manufacture.

Fixed shields or approved shields are required for:

Work with electric arcs or plasma arcs.

Processes where eye protectors or approved shields or approved fixed shields are required:

Gas welding.
Hot fettling of castings.
Flame cutting.
Lasers.
Truing or dressing of abrasive wheels.
Work with power hammers, drop hammers and horizontal forging machines.
Dry grinding of materials held by hand.
Fettling of metal castings.
Any machinery process where fragments may be thrown off.
Electric arc welding.

Protection required for persons in the presence of but not directly engaged in:

Chipping of metal or knocking out, cutting or shearing of rivets, bolts, plugs, nuts, lugs, pins, collars and similar articles by means of hand or portable power tools.
Processes where there are exposed electric or plasma arcs.
Work with drop hammers, power hammers or forging machines.
Fettling of castings.
Lasers.

Summary of Key Points

(1) Protective equipment should only be used when it has proven

impractical to control the hazard by engineering or other means.

(2) British Standards are available covering the majority of protective equipment specifications.

(3) Respiratory protection recommended practice is given in BS 4275.

(4) In order to ensure protection, a full programme must be set up including three key areas:

(a) administration;
(b) equipment (selection, use and maintenance);
(c) training.

(5) Responsibility for a protective equipment programme should be with line management with the occupational health unit as an integral part of the team.

Further Reading

ACGIH *Respiratory Protective Devices Manual* (ACGIH, Ann Arbor, Michigan, 1978)

British Standards Institute *British Standards Year Book* (BSI, London, annually)

Industrial Safety Manufacturers Association *Reference Book of Protective Equipment* (ISMA, London, 1978)

Riddel, F. 'Personal Protection and Equipment' in Handley, W. (ed.) *Industrial Safety Handbook*, Chapter 31 (McGraw-Hill, New York, 1977)

11 PLANNING SURVEYS AND USE OF RESULTS

Introduction

Industrial hygiene surveys require a broad knowledge of all aspects of the problem in hand. It is vital that the investigator recognises that evaluation of an industrial health hazard involves a number of areas, namely:

(1) Engineering details of the equipment.
(2) Toxicity of the materials.
(3) Process details – what reacts with what to give what.
(4) Production details – how many times a shift, etc.
(5) Medical aspects – worker history.

It is not possible for the investigator to be an expert in each of these fields, nor always to be able to enlist the required help. If information is *not* available, then the survey work carried out should take this into account. An overall checklist is appended to this chapter to allow the reader to ensure that the required areas are covered in the right order. The importance of doing things in the right order will become apparent as each step is covered below.

Planning the Strategy

Process and Operators

In order to set down a meaningful monitoring strategy, the investigator must know all about the process and the operators. This will involve: digging into the files to obtain the process charts; listing all chemicals used and produced; discussing the operators' function during the whole shift and determining from the engineers the potential 'leak' points, e.g. pumps, valves, drains and sampling points. Potential physical hazards should also be noted, i.e. heat, radiation, noise, etc. Generally at this stage the investigator will also visit the unit to confirm the points discovered so far.

Toxicity

Care must be taken to ensure that toxicity information is obtained on all chemicals used and produced. Sources are many for this kind of information – in particular the AIHA Hygiene Guide Series and the NIOSH *Toxic Substances.* (See Glossary for explanation of AIHA and NIOSH.)

With the toxicity data and the information on the physical agents present, the investigator can then define the areas of *potential hazard.* It must be remembered that although a chemical may be toxic, it is not a hazard until the operator is exposed to it.

Jobs Carried Out

Knowing the process, the chemicals and the toxicity information allows the investigator to examine the 'hazard potential' by linking these data with data on where the operators actually work. For example: an open plant such as a petrochemical complex may have areas where the atmospheric contaminant is at – or around – the short-term limit value. However, it is known that operators would only enter that unit area once per day for one or two minutes at most. In that situation the exposure would not be regarded as a hazard. If the work regime meant an operator had to work in that area for say two hours per day, then a hazard would exist and measures would have to be taken. In any event the investigator would make the plant management aware of the situation so that the correct hazard warning signs and training could be given to the operators to ensure that exposures were restricted to within acceptable limits.

The other important factor in this area is the overtime worked by the operators. Exposures that are not regarded as significant with a 40-hour working week may become so with excessive overtime. For example: a TLV (Threshold Limit Value) for 8-hour work-days of 100 ppm becomes 80 ppm for a 10-hour work-day. This *pro rata* method of working out extended TLVs is not accepted by some people but represents the simplest way of working out these limits.

What Methods of Control Are Already in Use?

If the process already incorporates a means of controlling exposure, then the investigator should include the testing of the effectiveness of this in the survey.

With these pieces of information the survey can now be planned.

The Survey

It is assumed that the investigator is aware of the limitations and accuracy of the equipment used and that the necessary calibrations have been carried out. (All equipment will have a designed range of conditions over which it can be used and may not be accurate outside of this range.)

Another factor that may affect the results however, is the variation in the exposure levels themselves. Even if the measuring equipment is 100 per cent accurate (it never is!) there will be variations in the true levels from day to day and maybe from minute to minute. Measurements should take this into account by covering several days (and nights if necessary) so that the natural variation can be measured. Open plants are particularly susceptible to this variation because of the effect of the wind and ambient temperature. If short-term variations are expected – say over 15 minutes or so – then the evaluation must include both short- and long-term readings.

Operators may also influence the way the survey is carried out. Almost inevitably the investigator will be using both static (area) sampling and personal (where the sampler is attached to the operator himself) sampling. Being human the operator may wish to test the system by holding his personal sampler near to a known source of contaminant in order to increase the reading. Although, as a conscientious investigator should always ensure, a briefing session will have been held with the operators to explain the reason for the survey and the need to obtain 'realistic' results, none the less false readings happen occasionally and the only guide that can be offered here is that they rarely happen *after* the first day or so of the survey. If high data appear on the first day but data of orders of magnitude lower after that, then call the operator in and discuss that day's operations. By working with him to discover the 'process' reason for the high reading, eventually it will emerge that he may have accidentally left it near a leaking pump (for example).

Very few operations are so consistent that less than four days of survey data are required to pinpoint accurately the average exposure *and* the peaks. Remember that the peaks may be just as important as the average.

As far as actually fixing the static sampling points is concerned, this will depend on the process and the investigator must take the layout into account. If personal samplers are being used, then the only purpose of static samplers is to determine where the source of contaminant lies (and for process control!) and knowledge of the process will define the points to site the samplers. The TLVs are for personal exposures, so whenever possible use samplers attached to the operator with the sampling head in the breathing zone. If this is not possible, then the static sampling should be increased and the personal levels estimated from these data and from the time the operator spends in each area (although this is not very satisfactory).

Planning surveys of this kind can be critical and if any doubts exist a specialist hygienist should be consulted.

Interpretation of Results

A great deal of commonsense and judgement needs to be used in interpreting the results of an environmental study.

The list of TLVs and STELs (Short Term Exposure Limit) given in the ACGIH annual recommendations are guidelines only to levels of exposure above which the hazards should be regarded as unacceptable. However, this does not mean they are target objectives; in fact, if it is reasonably practical to reduce exposure to zero then this should be done. Many statistical techniques exist to apply to data in order to check that exposure to above-TLV levels is unlikely to occur at a given probability. A better approach is to ensure that mean exposures are at most 50 per cent of the TLV with no excursions above the TLV, again applying a commonsense approach to ensure that the data are representative of the real situation.

Suppose, for example, that a survey is carried out of hydrocarbon exposure to shift personnel over a full 8-hour shift and the data are as follows:

	First Operator	Second Operator
Shift 1	10 ppm	21 ppm
Shift 2	5 ppm	27 ppm
Shift 3	0.1 ppm	27 ppm
Shift 4	0.9 ppm	25 ppm

Assume the TLV is 30 ppm with an STEL of 50 ppm, then these results show that the average reading for the first operator is 4 ppm and for the second operator 25 ppm. However, there is large variation for the first operator but less for the second. All data are below the TLV but the difference between the two operators should give cause for concern. The interpretation should be that there is a source of emission that could *potentially* give hazardous exposure levels and that the second operator seems to work in an area where this exists – although on some shifts the first operator may also be in this area, for shorter times.

The next step for the investigator is to determine where this area is, either from static sample data or from operator work procedures. Once discovered, the source should then be examined to determine if control is reasonably practical and action taken.

It is fair to say that this example is reasonably typical of industrial situations and the message is that TLVs are not absolute figures below which safety is guaranteed. Exposures which are atypically higher in certain areas should be examined to determine why and what remedial action is necessary.

The data from the survey may not be the only information available of course. Previous surveys may provide further comparative evidence of a worsening or improvement in exposure levels. Care must be taken to ensure that reporting of the data includes a description of the sampling and analytical method used. This avoids the situation of comparing two sets of data collected using different techniques. This must not be done except where information is available to confirm that the two methods are inter-changeable under the conditions used in the survey.

Hygiene Survey Checklist

(1) Arrange a meeting with management and the workforce in which the purpose of the survey is given. This meeting should include the full OH team where possible.

(2) Examine the plant operations/chemicals:
   2.1 Process charts.
   2.2 Inventory of raw materials and products.
   2.3 Toxicity data on all materials.
   2.4 Determine jobs and rotas carried out by operators.
   2.5 Discuss the medical status of workers with plant physician or OH nurse.
   2.6 Review previous studies of this plant.
   2.7 Walk through the plant to gain a subjective idea of potential problems.

(3) Study preparations:
   3.1 Decide on sampling and analytical techniques to measure

levels of contaminants.
3.2 Calibrate instruments.
3.3 Obtain protective equipment – if necessary.
3.4 Prepare written schedule of sampling and tables in which data can be logged easily.

(4) Carry out survey:
4.1 Advise management and workers of times of survey.
4.2 Confirm the work schedules unaltered.
4.3 Assemble the samplers (either personal or static) in the correct locations.
4.4 Record all relevant data:
(a) sample number;
(b) location/contaminant;
(c) time of start;
(d) flow-rate (check during run);
(e) time of end;
(f) comment on any unusual occurrence.
4.5 Remove sampling units.
4.6 Label for easy recognition later.

(5) Interpret data:
5.1 Calculate average.
5.2 Note lowest and highest levels.
5.3 Assess reliability of data.
5.4 Compare with TLV and STELs and from operator to operator, shift to shift.
5.5 Discuss results with both management and workforce.

(6) Recommend control measures – where necessary:
6.1 Engineering controls.
6.2 Administrative controls (work time, etc.).
6.3 Personal protection.
6.4 Medical surveillance of workforce.

Summary of Key Points

(1) Plan the strategy:

(a) Learn about process and operators.

(b) Obtain toxicity information.
(c) Examine work regime.
(d) Define existing controls.

(2) Prepare equipment carefully for easy use in the field:

(a) Calibrate beforehand.
(b) Use prepared tables for data logging.

(3) Record data/events as they happen – *do not* rely on memory.
(4) Interpret data against known standards but *do not* regard these as targets merely guidelines of acceptability.
(5) Discuss the results with management *and* workforce.
(6) Recommend control measures. These need not be just engineering measures. They could be administrative, relate to personal protection and they may include a medical surveillance programme for the relevant workers.

Further Reading

ACGIH 'TLVs for Chemical Substances and Physical Agents in the Workroom Environment with Intended Changes for 1980' (ACGIH, Cincinnati, 1980)

AIHA 'AIHA – Hygiene Guide Series' (AIHA, Ohio, issued as sheets regularly)

Health and Safety Executive 'Threshold Limit Values for 1979', Guidance Note EH 15/79 (HMSO, London, 1979)

NIOSH *Toxic Substances – Annual List* (US Government Printing Office, Washington, DC, annually)

12 BIOLOGICAL MONITORING

Introduction

Most air sampling techniques involve the operator wearing a personal monitor of some description, as near as possible to the breathing zone. Although this theoretically gives an indication of the contaminant *level* in the air *entering the lung,* it does not necessarily give a measure of the *amount absorbed by the body* in the long term. This will vary according to the particular absorption rate, the worker's breathing rate (varying with work rate) and metabolite exit rate. Arguably, therefore, the best way to measure an operator's exposure is to measure the level of the contaminant (or its metabolite) in the body or its excretions. The debate on this subject is complex and not yet resolved since it involves not only the validity of the data but also the setting of standards. If epidemiological evidence suggests that 10 ppm of substance X begins to have a physiological effect, then measurement of the 10 ppm in the workplace should be done by the same method as in the supportive work that set the standards.

Turning a Nelson-like eye to the debate, however, it is worthwhile considering some of the more common contaminant exposures that can be assessed from biological measurements. This is particularly important since this area is increasing in interest and the role and contribution of the nurse will be large in, for example, setting up the programme and obtaining samples.

Benzene

Benzene enters the body most easily via the lungs. It is partly metabolised to phenol which is excreted via the urine. Estimation of phenol in urine has, therefore, been used as a measure of benzene exposure. Phenol is always present from the diet however, usually providing up to 40 mg/1 in the urine. This therefore has to be used as a baseline. The difficulty arises since 10 ppm benzene in the air will create the same level, i.e. about 40 mg/1 of phenol in the urine. The technique is usually only useful at levels in excess of the TLV of 10 ppm.

Benzene is also excreted through breathing. Absorbed benzene migrates to the fatty tissue but once the exposure is stopped, i.e. at the end of the working day, the equilibrium is broken and the benzene begins to leave the fatty tissue and exit via the lungs. Benzene in the breath therefore is also a technique used to measure a previous shift exposure. Since these levels are in parts-per-billion the measurement technique to be used must be well-controlled. This technique is still under development but the reader is advised to follow its development in the journals since when fully correlated the technique will provide a simple means of obtaining an 8-hour personal exposure level – even for benzene levels in the air of less than 1 ppm.

Trichlorethylene (TCE)

Trichlorethylene is metabolised to trichloracetic acid (TCA) and excreted as such in the urine. This is probably the most widely employed biochemical parameter used for TCE exposure. A great deal of work has been done to correlate the TCE exposure with TCA in urine and an air concentration of 100 ppm can be related to a TCA excretion level of 100 mg/1 of urine.

Lead

Inorganic

Absorbed inorganic lead is stored in the skeleton and is also found in the blood. Excretion is via the urine and faeces. Because the lead also interferes with the haem-cycle enzymes (blood making) by inhibiting the formation of porphyrin prior to the coproporphyrin stage, the level of coproporphyrin III in the urine is increased.

There are three possible ways of assessing inorganic lead exposure, therefore:

(1) lead level in blood;
(2) lead in urine;
(3) coproporphyrin III in urine.

Although none is particularly satisfactory as daily monitoring devices, each can be used to check on the effect of previous periods of high exposure.

Lead Alkyls

Tetraethyl lead and tetramethyl lead have no effect on the haem cycle and are not generally stored in the body. Their effects are acute and biological monitoring is, therefore, of limited usefulness. In their draft 'The Control of Lead at Work Regulations – 1979' the HSE however recommend urinary lead measurements at regular intervals. Blood lead measurements are recommended at least once per year.

Mercury

Elemental and Inorganic

Elemental and inorganic mercury are excreted mainly via the urine and can be measured as such. One of the symptoms is proteinuria and therefore total protein in urine is also a guide to mercury exposure (although not specific to mercury).

Organo-mercury

This is also measured in the urine except methyl mercury which is mainly excreted via the faeces and should be measured there rather than in the urine. It should be noted that methyl mercury (causing Minamata Disease) passes the blood-brain barrier and placenta. Symptoms of the disease should also be looked for in this case.

Arsenic

This can be measured in the urine but also in the hair (pubic for preference) and nail clippings.

Cadmium

This is measured in the blood and urine.

Organo-phosphorus Compounds

These have a toxic action as a result of their inhibition of cholinesterase. Acetyl cholinesterase is the enzyme present at the nerve ending synapses. Measurement of cholinesterase in the blood can therefore be used as a guide to absorption of these compounds. The technique for measurement and the usefulness of this procedure are currently under debate. As a result other techniques have been considered. The measurement of nerve conduction velocity and electromyography (response of nerve to known electrical impulse) have gained in usefulness as they have been developed to indicate pre-clinical nerve changes. They do require expertise, however, and should only be carried out by experienced personnel.

Conclusion

This short discussion of several of the available techniques to provide biochemical information on exposure levels has served only as an introduction to the field. The debate as to the usefulness of pre-clinical biochemical monitoring will continue for some time as researchers gather more data. The reader is recommended to follow this debate closely as it could provide revolutionary changes to the work done by the occupational health team.

Summary of Key Points

(1) Atmospheric monitoring of breathing zone air may not be a good guide to levels of contaminant absorbed by the body in the long term, because of variations in:

- (a) breathing rate (work rate);
- (b) metabolite exit rate;
- (c) absorption rate in lungs.

(2) Analysis of the following can be used to gauge atmospheric exposure levels:

	Technique
Benzene	Phenol in urine; benzene in breath
Inorganic lead	Lead in blood/urine; coproporphyrin III in urine
Elemental mercury/ inorganic mercury	Mercury in urine; protein in urine
Methyl mercury	In faeces

Arsenic	In urine/hair/nails
Cadmium	In blood/urine
Trichlorethylene	In urine as trichloracetic acid
Organo-phosphorus compounds	Cholinesterase in blood/urine;* nerve conduction velocity; electromyography

* Some doubts about validity.

Further Reading

EMAS 'Occasional Paper I – Biochemical Criteria in Certain Biological Media for Selected Toxic Substances' (HMSO, London, 1974)

APPENDIX: LEGISLATION

In the preceding chapters the sections headed 'Legislation' have dealt with the specific legal requirements associated with the subject under consideration. In addition to these particular legal requirements it should be noted that the general requirements of the Factories Act 1961, the Offices, Shops and Railway Premises Act 1963, and the Health and Safety at Work Act 1974 have to be met.

Indeed the Health and Safety at Work Act (HASAWA) in its Section 1 deals specifically with the general requirement to secure the health, safety and welfare of persons at work and also to ensure that the health and safety of the community are not put at risk as a result of the activities of persons at work. In Section 2 of the Act, the employer is under obligation 'to ensure, so far as is reasonably practicable the health, safety and welfare at work of all his employees'.

The law therefore requires considerable efforts to be made in safeguarding the health and safety of the working and general populations and the reader is well advised to become familiar with the requirements of the Acts mentioned above. To assist in this the following summary of the sections of the HASAWA 1974 is given.

Summary of Health and Safety at Work Act 1974

Part I

The first part of the Act deals with health, safety and welfare in connection with work and control of dangerous substances and certain emissions into the atmosphere.

Sections 1-9. Deal with the preliminary and general duties of the employers and employees: safe plant, safe place of work, welfare at work, policy statements, safety representatives and duty to consult, duties of self-employed, emissions.

Sections 10-14. Deal with the establishment and functions of the Health and Safety Commission and Executive – establishing the commission committee composition, duties of control of emission, liaise with government departments, direct enquiries.

Sections 15-17. Deal with regulations and approved codes of practice.

Power to make regulations, exemptions, issue codes.

Sections 18-26. Deal with the enforcement of the Act and include information on the powers of inspectors, local authorities, specific powers of inspection-entry, inspectors' rights to take a constable on duty, investigate, samples, photograph issue notices.

Sections 27-28. Deal with the obtaining of information and its possible disclosure.

Sections 29-32. Deal with special provisions relating to agriculture, Agriculture Minister, regulations made by Minister of Agriculture, Fisheries and Food and Secretary of State.

Sections 33-42. Deal with offences under the Act, their punishment and persons to be punished: £400 maximum, two years imprisonment. Offences – individual guilt, inspectors to prosecute, what is practicable.

Section 43. Deals with the financial provisions of the Commission, etc. Funds dispersed by Secretary of State to the Committee to the Executive.

Sections 44-54. Deal with miscellaneous and supplementary items, appeals against licensing authorities, serving of notices, civil proceedings, metric units, meaning of work and at work, definitions.

Part II

Sections 55-60. Deal with the functions of the Employment Medical Advisory Service.

Part III

Sections 61-76. Deal with the Building Regulations and Amendment of Building (Scotland) Act 1959. Regulations for design, etc. of buildings, inspection, etc., planning approval, unsuitable materials, etc., tests for conformity.

Part IV

Sections 77-85. Deal with miscellaneous and general items, including Amendments to Radiological Protection Act 1970, to the Fire Precautions Act 1971, etc. Means of escape, Amendments of Companies Acts.

Further Reading on HASAWA 1974

Fife, I. and Machin, E. *Redgraves Health and Safety in Factories* (Butterworth, London, 1976, Supplement 1979)

Health and Safety Commission, Booklets – including: 'The Act Outlined', 'Advice to Employees', 'Advice to Employers', 'Advice to the Self-employed', 'Some Legal Aspects and How They Will Affect You' (HMSO, London, 1974)

HMSO *Health and Safety at Work etc. Act 1974* (HMSO, London, 1974)

Powell-Smith, V. *Questions and Answers on the Health and Safety at Work Act* (Alan Osborne & Associates (Books) Ltd, London, 1974)

Trades Union Congress *Safety and Health at Work* (TUC, London, 1978)

Wrigglesworth, F. and Earl, B. *A Guide to the Health and Safety at Work Act* (The Industrial Society, London, 1974)

LIST OF HYGIENE CONSULTANTS

The following are organisations offering industrial hygiene services on a fee-for-service basis. They will undertake measurements of worker exposure to chemical contaminants such as dust, fumes, vapours, gases, mists (oil etc.) and also to physical stresses such as noise, light, heat, etc.

(1) North of England Industrial Health Service,
20 Claremont Place,
Newcastle upon Tyne.
Contact Dr J. Steel (Tel. 0632-28511, ext. 3001)
(In association with the University of Newcastle upon Tyne)

(2) Environmental Health Service,
Level 5,
Medical School,
Ninewells,
Dundee,
Scotland.
Contact T. Gillanders (Tel. 0382-644625)
(In association with the University of Dundee)

(3) National Occupational Hygiene Service Ltd,
12 Brook Road,
Fallowfield,
Manchester 14.
Contact Mr E. King (Tel. 061-224-2332)
(Formerly in association with the University of Manchester)

(4) TUC Centenary Institute of Occupational Health,
London School of Hygiene and Tropical Medicine,
Keppel Street (Gower Street),
London WC1.
Contact Director (Tel. 01-636 8636)
(In association with the University of London)

(5) Occupational Health and Safety Group,
Department of Epidemiology and Community Medicine,
Welsh National School of Medicine,
Heath Park, Cardiff CF4 4XN.
Contact Mr A. Samuel (Tel. 0222-755944)
(In association with the University of Wales)

LIST OF SUPPLIERS OF SAMPLING EQUIPMENT

This should not be regarded as an exhaustive list of equipment suppliers but is merely given to 'help you on your way'.

Dust Sampling Pumps

C.F. Casella & Co. Ltd.
Regent House,
Britannia Walk,
London N1 7ND

Rotheroe & Mitchell Ltd,
Victoria Road,
Ruislip,
Middx HA4 1LG

Vapour/Gas Sampling Pumps

As above – plus:

MDA Scientific (UK) Ltd,
Ferndown Industrial Estate,
86 Cobham Road,
Wimborne,
Dorset BH22 7PQ

Draeger Safety,
Draeger House,
Sunnyside Road,
Chesham,
Bucks.

Direct Reading Instruments

Draeger Safety,
Draeger House,
Sunnyside Road,
Chesham,
Bucks.

D.A. Pitman Ltd,
Jessamy Road,
Weybridge,
Surrey

Vertec Scientific,
11 Hay Hill Rise,
Taplow,
Nr Maidenhead,
Berks.

Dust/Gases/Vapour Sampling Heads

C.F. Casella & Co. Ltd,
Regent House,
Britannia Walk,
London N1 7ND

Gelman Hawksley Ltd,
10 Harrow Road,
Brackmills,
Northampton NN4 0EB

Dust/Gases/Vapour Sampling Heads (contd.)

Environmental Monitoring
Systems,
Marketed by:
Bastock Marketing,
5 Portway Gardens, Aynho,
Banbury, Oxon. OX17 3AR

Noise Equipment

B & K Laboratories,
Cross Lances Road,
Hounslow,
Middx TW3 2AE

Castle Associates,
Redbourn House,
North Street,
Scarborough YO11 1DE

Computer Engineering Ltd,
14 Wallace Way,
Hitchin,
Herts. SG4 0SE

RECOMMENDED FURTHER READING

Chemical Society *Hazards in the Chemical Laboratory* (The Chemical Society, London, 1977)

Health and Safety Executive, Various Guidance Notes (HMSO, London)

Hunter, D. *The Diseases of Occupations* (The English Universities Press, London, 1975)

International Labour Office *Encyclopaedia of Occupational Health and Safety* (ILO, Geneva, 1974)

NIOSH *The Industrial Environment – its Evaluation and Control* (US Government Printing Office, Washington, DC, 1973)

—— *Occupational Diseases – A Guide to Their Recognition* (US Government Printing Office, Washington, DC, 1977)

Patty, F.A. *Industrial Hygiene and Toxicology*, 3 vols (John Wiley, New York, 1978)

Plunkett, E.R. *Handbook of Industrial Toxicology* (Chemical Publishing Co., New York, 1976)

Sax, N. *Dangerous Properties of Industrial Materials* (Van Nostrand Reinhold Co., New York, 1979)

Trades Union Congress *Health and Safety at Work – A TUC Guide* (TUC, London, 1978)

Trevethick, R.A. *Environmental and Industrial Health Hazards – A Practical Guide* (William Heinemann Medical Books, London, 1976)

Waldron, H.A. and Harrington, J.M. (1981) *Occupational Hygiene: an Introductory Text* (Blackwell Scientific Publications, Oxford)

GLOSSARY OF TERMS AND ABBREVIATIONS

Absorption: Penetration of a substance into the body of another.
ACGIH: American Conference of Governmental Industrial Hygienists.
Acoustic: Associated with sound.
Acoustic trauma: Hearing loss caused by a sudden loud noise in one ear or a sudden blow to the head. Hearing loss is usually temporary although there may be permanent loss in some cases.
Acuity: This pertains to the sensitivity of hearing.
Auditory acuity: Ability to hear clearly and distinctly.
Acute: Short and severe, not prolonged or repeated as for chronic. Its use in the context 'acute toxicity' denotes the toxicity of a substance or physical agent, produced by a brief or single exposure of living things (man, animals, etc.), to the substance or agent.
Aerosols: Liquid droplets or solid particles dispersed in air, that are of fine enough particle size (0.01 to 100 micrometres) to remain so dispersed for a period of time.
AIHA: American Industrial Hygiene Association.
Air monitoring: The sampling for and measurement of pollutants in the atmosphere.
Albuminuria: The presence of albumen or other protein substances such as serum globulin in the urine.
Aliphatic: Pertaining to an open-chain carbon compound. Mainly applied to products derived from a paraffin base and having straight or branched chain structures.
Allergy: An exaggerated susceptibility to particular foreign substances or physical agents which have no effect with most individuals.
Alveoli: Tiny air sacs of the lungs, found at the end of each bronchiole. Through the thin walls of the alveoli the blood takes up oxygen from the air and gives up carbon dioxide which is breathed out.
Amorphous: Literally 'of no definite shape' or non-crystalline as in the case of one form of silica for example.
Anaemia: A disorder of the blood due to a deficiency in the number of red blood cells or their oxygen carrying component, haemoglobin.
Anemometer: An instrument for measuring the force or velocity of air.
Anaesthesia: Loss of sensation, in particular the temporary loss of feeling, induced by certain chemical agents.
Anions: Negatively-charged ions.

Aplastic anaemia: A condition in which the red bone marrow is unable to produce the normal supply of red blood corpuscles.

Aromatic: Applied to a group of hydrocarbons characterised by the presence of a benzene ring.

Asbestosis: A disease of the lungs caused by the inhalation of fibres of asbestos.

Asphyxia: Simple asphyxia is suffocation from a lack of oxygen supply. Chemical asphyxia is produced by a substance, such as carbon monoxide, that combines with haemoglobin to reduce the blood's capacity to transport oxygen.

Asthma: Condition involving difficulty in breathing due to the spasmodic constriction of the bronchial muscles often in response to irritants, allergens or other stimuli.

Atoms: Smallest unit recognised as a particular element.

Bactericide: An agent that destroys bacteria.

Benign: Not malignant. A benign tumour is one which is not invasive nor does it metastasise.

Blood: Consists of many elements, including red and white blood cells, platelets and plasma. It transfers oxygen and carbon dioxide between the lungs and body cells but also carries nutrients that cells need and gets rid of their waste products via the kidneys and other excretory organs.

Blood count: The number of corpuscles per cubic millimetre of blood. Separate counts may be made of red and white corpuscles.

BOHS: British Occupational Hygiene Society.

Bone marrow: A soft tissue which constitutes the central filling of many bones. Some bone marrow, known as red marrow, serves to produce blood cells.

Bronchiole: The thinnest of the tubes that carry air into and out of the lungs.

Bronchitis: Inflammation of the bronchi or bronchial tubes.

BS: British Standard.

Carcinogenic: Producing cancer.

Carcinoma: Malignant tumour derived from the epithelial tissues, e.g. of the skin and membranes lining the body cavities and from certain glandular tissue.

Cataract: An opaque body which forms in the eye obscuring the transparency of the lens.

Cations: Positively-charged ions.

Cell: The basic unit of which body tissues are made.

Ceiling values: Values that must not be exceeded at any time as in levels of airborne contaminant.

CET: Corrected effective temperature.

Chloracne: Caused by chlorinated organic compounds acting on the sebaceous glands and the liver.

Chronic: Prolonged or repeated. 'Chronic toxicity' is the toxicity of a substance or physical agent produced by repeated or prolonged exposure to the substance or agent.

Cilia: Tiny hair-like projections from certain epithelial cells. Membranes containing such cells are known as ciliated membranes and can be found for example in the respiratory tract. The wave-like motion of these cilia clears foreign particles from the lungs by pushing them back up the system.

Contact dermatitis: Dermatitis caused by a *primary* irritant.

Cornea: Transparent membrane covering the outer portion of the eye.

Corpuscle: A red or white blood cell.

Cyanosis: Blue appearance of the skin especially on the face and extremities indicating a lack of sufficient oxygen in the arterial blood.

Decibel (dB): Prime unit for sound measurement.

Dermatitis: Inflammation of the skin. There are two general types of skin reaction, primary irritation dermatitis and sensitisation dermatitis (see irritant and sensitiser).

Diuretic: Any substance that promotes the excretion of urine.

Dyspnea: Shortness of breath, laboured breathing.

Eczema: A skin disease or disorder. A dermatitis.

Edema: A swelling of body tissues as a result of being waterlogged with fluid.

EEC: European Economic Community.

Element: A material which cannot be divided into separate chemical entities.

Embryo: The term applied to the developing ovum from conception to the third month of pregnancy.

Electrons: Negatively-charged particles.

Emphysema: A lung disease in which the walls of the alveoli have lost their elasticity.

Enzymes: Chemical substances, based on proteins, that promote chemical reactions in living organisms.

Epidemiology: The study of the relationships of the factors determining the frequency and distribution of diseases in a community.

Epidermis: The superficial strata of the skin.

Erythema: Reddening of the skin.

ET: Effective temperature.

Etiology: A study or knowledge of the causes of disease.

Fainting: Known as syncope, it is a temporary loss of consciousness as a result of a diminished supply of blood to the brain.

Fibrosis: A condition marked by an increase of interstitial fibrous tissue.

Fume: Particles of airborne dust which are less than 1 micrometer in diameter.

Fume fever: Metal fume fever is an acute condition caused by brief excessive exposure to the fumes of metals, such as zinc, copper or magnesium.

Gas: Material which is in the gaseous phase under ambient conditions.

Gingivitis: Inflammation of the gums.

Gonads: The male (testes) and female (ovaries) sex glands.

Haematology: The study of the blood and blood-forming organs.

Haematuria: Blood in the urine.

Haemoglobin: The material in the blood which carries oxygen. When oxygenated this gives the blood its brilliant red colour.

Haemolysis: Breakdown of red blood cells as a result of the cell wall bursting so releasing the haemoglobin.

Haemoptysis: Bleeding from the lungs, spitting blood or blood-stained sputum.

Haemorrhage: Bleeding, especially profuse bleeding, as from a ruptured or cut blood vessel.

Hepatitis: Inflammation of the liver.

Hertz (Hz): Frequency of sound (formerly cycles per second).

Hydromeiosis: Failure of function of sweat glands.

HSI: Heat stress index.

ILO: International Labour Office.

Ingestion: The process of taking substances into the body, as food, drink, medicine, etc. via the gastro-intestinal tract.

Innocuous: Harmless.

Inorganic chemistry: The study of the chemistry of the elements other than carbon.

Intoxication: Drunkenness or poisoning.

Ion: Electrically-charged atom which has gained or lost one or more electrons.

Ionising radiation: Part of the electromagnetic spectrum of energy that can cause chemical change by ionising molecules.

Irritant: A primary irritant is an agent that produces irritation (e.g. on the skin).

Jaundice: A symptom of disease that causes the skin, the whites of the eyes and even the mucuous membranes to turn yellow.

Keratitis: Inflammation of the cornea with tiny lesions.

Keratogenic and neoplastic agents: Agents which stimulate abnormal growth in the outer layer of skin and may cause tumours.

Laryngitis: Inflammation of the larynx.

Larynx: The organ by means of which the voice is produced. It is situated at the upper part of the trachea.

LD50 (Lethal Dose 50): The dose which produces death in 50 per cent of the exposed species. (A term used to define the toxicity of a material.)

Lethal: Capable of causing death.

Leukaemia: A blood disease distinguished by over-production of white blood cells. Almost always fatal.

Leucocyte: White blood cell.

Leucocytosis: An abnormal increase in the number of white blood cells.

Leucopenia: An abnormal decrease in the number of white blood cells.

Liver: The largest organ in the body, situated on the right side of the upper part of the abdomen. It has many important functions such as regulating the amino acids in the blood, forming and secreting bile which aids in absorption and digestion of fats, transforming glucose into glycogen, etc.

Localised: Restricted to one spot or area in the body and not spread all through it – contrast with systemic.

Lux: The illuminance per square metre of surface area.

Malignant: As applied to a tumour – cancerous and capable of undergoing metastasis.

Metabolism: Describes the changes that take place in the cells that make up the living body.

Miliary: Characterised or accompanied by seed-like blisters or inflamed, raised portions of tissue.

Molecules: The smallest individual part of a compound.

MMMF: Man-made mineral fibres.

Narcosis: Stupor or unconsciousness produced by chemical substances.

Narcotics: Chemical agents that induce narcosis.

Necrosis: Destruction and death of body tissue.

Nephritis: Inflammation of the kidneys.

Neuritis: Inflammation of a nerve.

Neutron: A particle with no electrical charge.

NIOSH: National Institute of Occupational Safety and Health (USA).

Non-ionising radiation: Part of the electromagnetic spectrum that does *not* have sufficient energy to cause ionisation of molecules.

NPF: Nominal protection factor.

OH: Occupational health.

Organ: An organised collection of tissues that has a special and recognised function.

Organic chemistry: Study of the chemistry of carbon. Many of the compounds are found in nature.

Osseous: Pertaining to bone.

Overexposure: Exposure beyond the specified limits.

Pathology: The study of disease processes.

Peripheral neuropathy: An effect on the nervous system of the body extremities.

Personal sampling: Sampler attached to operator.

Physiology: The science and study of the functions or actions of living organisms:

Plasma: The fluid part of the blood in which the blood cells are suspended.

Pneumoconiosis: Disease of the lungs caused by the excessive inhalation of various kinds of dusts and other particles.

Pneumonitis: Inflammation of the lungs characterised by an excess of fluid in the lungs.

PPM: Part per million.

Prophylactic: Preventive treatment for protection against disease.

Proteins: Large molecules found in the cells of all animal and vegetable matter and containing carbon, hydrogen, nitrogen and oxygen, sometimes sulphur and phosphorus.

Protons: Positively-charged particles.

Purpura: Extensive haemorrhage into the skin or mucous membrane.

Retina: Light-sensitive inner surface of the eye which receives and transmits the image formed by the lens.

Rhinitis: Inflammation of the mucous membranes lining the nasal passages (usually resulting in runny nose).

Sclera: Hardening of the eyeball.

Scleroderma: Hardening of the skin.

Sensitiser: Substance or agent which initiates production of a defensive mechanism within the body which is triggered to produce an allergic type response on second or subsequent exposures.

Septicaemia: Blood poisoning.

Silicosis: A disease of the lungs caused by the excessive inhalation of crystalline silica dust.

Skin: Thin layer of tissue covering the outer surface of the body, divided into two main layers, the superficial layer known as the epithelium, epidermis or cuticle, and the deeper layer known as the corium or derma.

Spasm: Tightening or contraction of any set of muscles.

Static sampling: Fixed at specific distance.

Stroke: Any sudden, severe attack of illness, usually associated with brain haemorrhage.

Sub-acute: An illness or condition that is not quite as serious or as dangerous as the acute phase.

Symptom: Any evidence from a patient that he is sick.

Systemic: Affecting the body as a whole rather than just a local spot.

TEL: Tetraethyl lead.

Thermocouple: A device comprising two dissimilar metals formed together at two points whereby a temperature difference between these points induces an electric current to flow.

Threshold: The level where the final effects occur.

Threshold Limit Value (TLV): The airborne concentration of a substance to which it is believed that nearly all workers may be repeatedly exposed day after day without adverse effect. There are three categories:

Time-weighted Average (TLV-TWA): The time-weighted average concentration for a normal 8-hour workday or 40-hour week to which nearly all workers may be repeatedly exposed, day after day, without adverse effect. This is the most common category. It should be used as a guide only and not as a fine distinction between a safe and dangerous situation.

Short-term Exposure Limit (TLV-STEL): The maximum concentration to which workers can be exposed for a period up to 15 minutes *continuously* without suffering adverse effects.

Ceiling (TLV-C): The concentration that should not be exceeded even *instantaneously.*

TML: Tetramethyl lead.

Toxaemia: Poisoning by way of the bloodstream.

Toxicity: The inherent potential of a material to cause an adverse effect on biological systems.

Trachea: The windpipe or tube that conducts air to and from the lungs. It extends between the larynx and the point where it divides into two bronchi.

Tremor: Involuntary shaking, trembling or shivering.

UV Ultra-violet.

Vapour: Evaporated state of materials which are liquid at ambient temperature and pressures.

Velocity: Rate of motion.

Vertigo: The sensation that the environment is revolving around you.

Vesicle: A small blister on the skin.

Virulent: Extremely poisonous or venomous; capable of overcoming bodily defensive mechanisms.

WHO: World Health Organisation.

WBGT: Wet-bulb globe temperature.

INDEX